T0318638

Mechatronic Components

Mechatronic Components

Roadmap to Design

Emin Faruk Kececi

Butterworth-Heinemann is an imprint of Elsevier
The Boulevard, Langford Lane, Kidlington, Oxford OX5 1GB, United Kingdom
50 Hampshire Street, 5th Floor, Cambridge, MA 02139, United States

Library of Congress Cataloging-in-Publication Data
A catalog record for this book is available from the Library of Congress

British Library Cataloguing-in-Publication Data
A catalogue record for this book is available from the British Library

ISBN: 978-0-12-814126-7

For information on all Butterworth-Heinemann publications
visit our website at https://www.elsevier.com/books-and-journals

Publisher: Mara Conner
Acquisition Editor: Sonnini Yura
Editorial Project Manager: Naomi Robertson
Production Project Manager: Sruthi Satheesh
Cover Designer: Mark Rogers

Typeset by SPi Global, India

Contents

Preface

The different components of the mechatronic system are already explained in field-specific books but since mechatronics is an interdisciplinary subject of mechanical, electrical, and computer engineering, it is hard for an engineer to understand all the different components of these fields. It is especially important to understand the different choices of components and how they are different from each other.

In this book, approximately 130 components are explained with diagrams and flowcharts, so that any engineer from the aforementioned fields can understand the general properties and selection criteria of a component. It is considered that this book will be useful to upper-level undergraduate students, graduate students, and professional engineers to help them to understand the different components used in their mechatronics projects (mostly mechanical, electrical, and computer engineering majors).

Acknowledgments

"It takes a village to raise a child" is a beautiful African proverb to explain the process of writing a book. There were so many people who have helped me in one way or the other to put this book together and I am very grateful for their kindness, hard work, and support.

Elsevier Publishing Company is so professional that in almost 2 years after I proposed this book it is finding itself on the bookshelf. It was a pleasure to work with Sonnini Yura and Naomi Robertson. I was very pleased with their promptness and support. They made the process so smooth that I would love to work with Elsevier again for my next book project.

The students of MKM 503E Mechatronic System Components course at Istanbul Technical University helped me to develop these ideas. It is due to their hard work that this book came together as it is.

I would also like to thank my doctoral student, Majid Mohammad Sadeghi, whose wondering mind always surprises me and who gave me ideas about this book and criticized me the most, which I hope made the book better.

Finally, I would like to thank my family: my daughter Ann Ayse Yasam Kececi, my son Yasar Daniel Kececi for their unlimited love, and my wife Anne Adams Kececi for her patience and support.

I would like to dedicate this book to my father and my mother, who taught me the importance of education and provided me the opportunity.

October 11, 2018
Istanbul, Turkey

Chapter 1

Mechatronics: A Brief History

Ten years ago in 2008, an article was published in the journal *Control Engineering Practice* titled as "Mechatronic systems—Innovative products with embedded control" by Rolf Isermann from Darmstadt University of Technology, Germany. This article describes the mechatronics as "Many technical processes and products in the area of mechanical and electrical engineering are showing an increasing integration of mechanics with digital electronics and information processing. This integration is between the components (hardware) and the information-driven functions (software), resulting in integrated systems called mechatronic systems."

After 10 years when the cited documents were examined, there were 173 of them. When we look at the cited documents, we can see titles such as "Automotive control: the state of the art and perspective," "The Internet of Things—The future or the end of mechatronics," "Multisensor fusion and integration: A review on approaches and its applications in mechatronics," and "Beyond advanced mechatronics: new design challenges of Social-Cyber-Physical systems."

As we can see from this simple example mechatronics is everywhere, from automotive control to internet of things to cyber-physical systems. So, one should wonder what happened that the concepts of mechatronics was able to grow this fast in the last decade. Because when we consider mechanical engineering, this field has been around for thousands of years and gained momentum with the Industrial Revolution. Same growth happened with electrical engineering which started to develop in the 19th century and still continues to advance.

Mechatronics is not a new concept, it is not even a field; it is a way of thinking and natural way of growth of mechanical engineering. As mechanical engineers we have realized that so many things inside the machines can be achieved by using electronics much more easily rather than making the whole system all mechanical. Moreover, the importance of both control and measurements greatly advanced the machines and all these necessities helped the field of mechatronics to grow.

When we look at the last decade we can understand how mechatronics systems are developing with an increasing speed. There are six reasons for this exponential growth. They are as follows:

1. *Internet access:* With the ease of internet access it is now much easier to access the information. This information what the user is looking for can be a

Mechatronic Components. https://doi.org/10.1016/B978-0-12-814126-7.00001-3

motor supplier, data sheet of a sensor, or some theoretical knowledge. Before the ease of internet access main source of knowledge were books. Considering that it takes at least 3–4 years for a book to start as an idea and finds itself a space on a bookshelf, the information received from a book is actually old knowledge in today's world. It is much easier to update information on websites and users can learn much faster.

2. *Open Source Communities*: Use of knowledge and sharing of the knowledge led people to form open source communities where people put their project information and even codes on the internet. When another person picks up the project and develops the codes the result can be impressive. RepRap Community worked together and filament-based three-dimensional (3D) printing technology has been developed. With the same intentions, Arduino Project was made public and all kinds of Arduino compatible electronic parts are developed.

3. *Ease of Manufacturing*: Manufacturing of the parts is getting easier. The computer-aided drawing programs were very expensive and very difficult to learn. Nowadays, you can download freeware drawing programs and even some of these programs are working online making it easier for the user by not needing a high computing power computer. Since people can draw the part they want easily, there are also online manufacturing companies where one can order a part to be manufactured by computer numerical control (CNC) laser cutting or CNC milling.

Especially the growth of 3D printing machine market and a desktop 3D printer getting cheaper makes prototyping of the machines much easier. Moreover, the availability of online 3D printing services makes even non-engineers able to design and build their ideas.

4. *Maker Spaces*: The necessity of hands-on experience in the engineering education is even more important as the competitions between the companies are increasing and for any start-up to be successful, they should be able to show their prototype. In order to let the engineering students to have hands-on experience, understand the theoretical part of the courses they learn, and learn to work in a team, different universities have Maker Spaces where the student can build their projects. This building and do it yourself culture are growing and closing the gaps between the engineering fields, especially when a machine is being built mechanical, electrical, and computer engineering students work collaboratively and understand the importance of mechatronics. In their professional life, they have also very big tendency to work together with the engineers from other disciplines.

5. *Off-the-shelf Electronics*: Embedded systems are very important part of mechatronic systems, where the software is put to run the system. The microcontrollers have been around, a user can buy the microcontroller but has to develop the board himself/herself. This process was very time consuming and since the board was not done professionally it was susceptible to mistakes. Off-the-shelf electronics overcame this problem and provided the users with ready-to-go programming board. It was revolutionary to have a board which

needed no soldering, could be powered by a USB port, programmed by a USB port (not needing to transfer the microcontroller to the programmer and back to the board), and provide terminal to see the inputs and outputs in real time. Arduino was the first commercially successful project and others have followed. The programming cards now come with different computing power, size, and input/output ports for different purposes. Sensor boards, different types of motor drivers, and all kinds of boards followed this trend.

It is naturally more expensive to use off-the-shelf electronics, but for educational and prototyping purposes use of these electronics cuts the time drastically.

6. *Cost Drop of Electronics and Manufacturing*: Around 15 years ago, inertial measurement unit (IMU) sensor from Honeywell was priced around 1000$. Nowadays, price of an IMU sensor from Sparkfun already on the board using I2C protocol ranges from 10 to 35$. It is also very easy to use online 3D printing companies or university services to build parts by using 3D printing technology where a gram of printed material cost 0.03$, and for an hour of printing the cost is 5$. The drop in the cost of electronics and 3D manufacturing is allowing more people to work on mechatronic projects.

When we consider a mechatronic system, it consists of mechanical, electrical, and software subsystems. This division in fact also shows itself in the literature. If the author of the book has a mechanical or electrical background, the book is also written from that perspective. This is the unique purpose of this book because people from different background do not have enough information to understand where to start to solve their problems in mechatronics.

Consider a mechanical engineer who needs to design a transmission system. There are many choices, such as gears, belts, chains, and shafts. In mechanical engineering education these concepts are taught in detail. However, if an engineer with an electrical engineering background is working on a transmission system, it is very time consuming even to realize the difference between different choices.

This book does not intend to teach the user different components of the mechatronics systems, rather it is a guidebook for the users to understand the advantages and disadvantages of different components.

This book consists of 16 other chapters. Chapter 2 describes the use of this book. Chapters 3–15 show with block diagrams different components and their usages and their kinds. These components are grouped as follows:

Chapter 3. Calculations of Mechanical Properties
Chapter 4. Mechanical Failure Modes
Chapter 5. Materials Properties
Chapter 6. Manufacturing Processes
Chapter 7. Machine Elements
Chapter 8. Design and Analysis Programs
Chapter 9. Assembly Processes
Chapter 10. Electronic Components

Chapter 11. Actuators
Chapter 12. Sensors
Chapter 13. Signal Processing
Chapter 14. Controls Theory and Applications
Chapter 15. Design and Simulation Softwares.

Finally, in Chapter 16 case studies are presented with examples to explain the reader the use of the book.

The idea of this book started around 2002, when the author read the book "The Practice of Machine Design" by Yotaro Hatamura, ISBN 978-0198565604. In this book some of the mechanical engineering concepts were explained by using tables and charts, making it very easy to understand the difference between the different concepts.

Chapter 2

Use of This Book: Mechatronic Components: Roadmap to Design

Design concept requires creativity and knowledge. The designer is presented with a vaguely defined problem. He will need to imagine a solution and use the engineering knowledge to design and build the prototype system.

Almost all the time the problem can be explained in one or two sentences, mostly by nonengineers. Let us say that the problem is to have a device to harvest energy from the water flowing in a creek. The user lives in a remote area and would like to have energy source for his home. From the user's perspective the problem is quite simple, but from the mechatronic engineer's point of view the limitations of the problem should be defined (the boundary conditions). This means that the designer should know the speed of the water, the depth and width of the creek, hourly need of the power requirements of the house, and energy storage system requirements for the times when the system is not in use. This engineering definition is very important and sets base for the calculation of the system. When the necessary calculations are carried out the designer can find the properties of the parts that should be used.

Since the concept of mechatronics is so wide, with the possibilities of different machines for different purposes, there are many choices in front of the designer to choose from. Let us consider transmission of rotational energy from one point to another. River flow energy harvester, Fig. 1, has a wheel which is rotated by the flow of the water and this rotation is transmitted to the electric generator shaft via belt system. However, for the transmitting of rotation from one shaft to another, the designer can use a belt, a chain, or a shaft system. Each of these different solutions is valid, but they naturally have advantages and disadvantages.

The purpose of this book, *Mechatronic Components: Roadmap to Design*, is to help a designer to choose from different subsolutions to solve his design problem. This book is basically a guidebook where the reader can learn different types of transmission systems and can decide which one to choose. Especially all the concepts are explained via block diagrams so that the designer can follow the system requirements to find the necessary solutions.

Mechatronic Components. https://doi.org/10.1016/B978-0-12-814126-7.00002-5

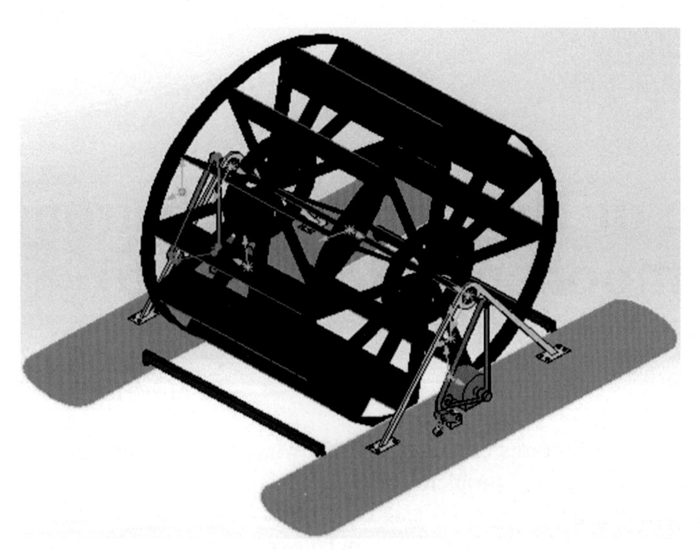

FIG. 1 Energy harvester from river flow.

Considering that there are mechanical, electrical, and computer scientists as well as people from other engineering fields working on mechatronic design projects, it is not quite common for the mechanical engineers to know much about communication protocols such as I2C, or for the electrical engineers to know the different types of three-dimensional (3D) printer materials. This book aims to guide engineers to find different possibilities for their problems. The user of course should learn the details of the systems, such as the working principle of I2C communication protocol, but this book explains the concepts from the point of view of the system requirements.

As the reviewers of this book stated, the audience of this project are definitely professional engineers, who sometimes need a reference handbook to quickly find some properties of the listed components.

"Mechatronics is an active and growing field. As the research and the technology advance, there are stronger and stronger requirements to attain better and better design process and performance. This can only be achieved through a complication of the system and a better exploitation of its different components, intended as integrated parts of the whole. This evolution generates needs for specific design tools. This book enters this phase of the process, by providing the technological and scientific community with a novel, rationale tool."

"Mechatronics is one among the multidisciplinary branch of learning that is gaining more and more space within the current scientific community. Intended as the combination of mechanics and electronics, including control systems and other related features, it is a huge domain that can be assumed to integrate other emerging fields like adaptive structures and robotics. The complexity of those subjects is the integration of different knowledge, coming from very different experiences.

The main difficulty is therefore to define a correct design approach that, without a guide, can be very chaotic and confused (for objective reasons). Any kind of assistance in such a definition is welcome and, currently, I can count only few trials in this direction. This is the reason why I believe this book goes in the right direction and promises to give a relevant, though preliminary, contribution."

In the academic environment especially students coming from very different backgrounds. Using this book will guide them what to study and in which path to start to learn more so that they can make their projects and achieve their research. In academia, as mentioned by the reviewers in the third section, this book can be used for capstone design courses, graduate level courses, and in project-based courses.

This book will also be of interest to researchers and professionals involved with the practical application of mechatronics, including specialists in the adaptive structures, robotics, and other fields where mechanics and electronics are jointly involved. PhD and graduate students involved in similar themes can also take advantage of this textbook for its large use of flow diagrams and schemes, helping to understand the logic behind the design of such systems. In this sense, it could be an excellent education book.

The primary audience would be practitioners in robotics, mechatronics, instrumentation, and control systems. Graduate students and researchers in these areas might also find this book handy as a reference book. The book could also be used as a reference book by upper-level undergrads for capstone design courses.

This book designed to offer detailed guidance in the selection of materials and components for building mechatronic systems. Compared with other mechatronic books, this book is likely going to be unique in that (1) it offers guidance on practical development of mechatronic systems with detailed information that is often unavailable in other books; and (2) it provides flow charts that guide the decision process for the user.

In order to use this book effectively, the designer should familiarize himself with the different chapters, where mechanical, electrical, and programming concepts are grouped. When the designer knows what type of problem he should solve, such as selection of motors, then he can look at the related section to see his options and advantages and disadvantages.

Chapter 3

Calculation for Mechanical Properties

ABSTRACT

During the design process, the machines work with basic principles of motion. When there is a motion, there is always friction and the phenomenon needs to be handled: (1) by reducing it as much as possible by using different kinds of bearings, mentioned in the mechanical element chapter, and (2) by calculating the required additional energy to overcome the frictional losses. The center of gravity can be calculated via the design software very easily, but the mechatronic designer should understand the concept and have a sense of its location during the design process. The center of gravity can determine how some forces, such as centrifugal forces, will form during the motion of the system. The moment of inertia is the resistance of the part to rotation and can be calculated depending on the geometry of the part. Without using more material, it is possible to have a part that has a higher moment of inertia. Thus, the designer can have a lightweight design with a higher strength. The Reynolds Number defines the flow characteristic: laminar or turbulent. The flow rate can be important if the designer would like to calculate the energy lost due to the friction inside a pipe. It is very important to understand the calculation formulas for these different kinds of properties, because the designer can understand the correlation between the design parameters by studying these formulas and thus make a correct decision to change the necessary parameters. It should be also kept in mind that for a given part, more than one property can be important and the parameters might be optimized at the same time. For example, changing the geometry of the part can change both the moment of inertia and the center of gravity.

Mechatronic Components. https://doi.org/10.1016/B978-0-12-814126-7.00003-7

Calculation for Friction

Friction causes a force resisting the relative motion of solid surfaces, fluid layers, and material elements sliding against each other.

Friction environment

Solid? Yes / No

Fluid? Yes / No

Gas

Input	Formula	Output
Static objects: $\mu_s = static$ $friction\ coeff$ $N = normal\ force$ $to\ friction\ surface$	$F = \mu_s \times N$	$F = friction\ force$
Sliding objects: $\mu_k = kinetic$ $friction\ coeff$	$F = \mu_k \times N$	
Rolling objects: $u_r = rolling$ $friction\ coeff$ $W = weight$	$F = u_r \times W$	
$\rho = fluid\ density$ $u = flow\ velocity$ $C_D = drag\ coeff$ $A = reference\ area$	$F_D = \frac{1}{2} \times \rho \times u^2 \times C_D \times A$	$F_d = drag\ force$
$\mu = viscosity\ coeff$ $A = reference\ area$ $u = velocity\ of$ $top\ layer$ $y = fluid\ depth$	OR $F = \mu \times A \times \frac{u}{y}$	$F = viscosity$
$\rho = gas\ density$ $u = flow\ velocity$ $C_D = drag\ coeff$ $A = reference\ area$	$F_D = \frac{1}{2} \times \rho \times u^2 \times C_D \times A$	$F_D = drag\ force$

Calculation for Center of Gravity

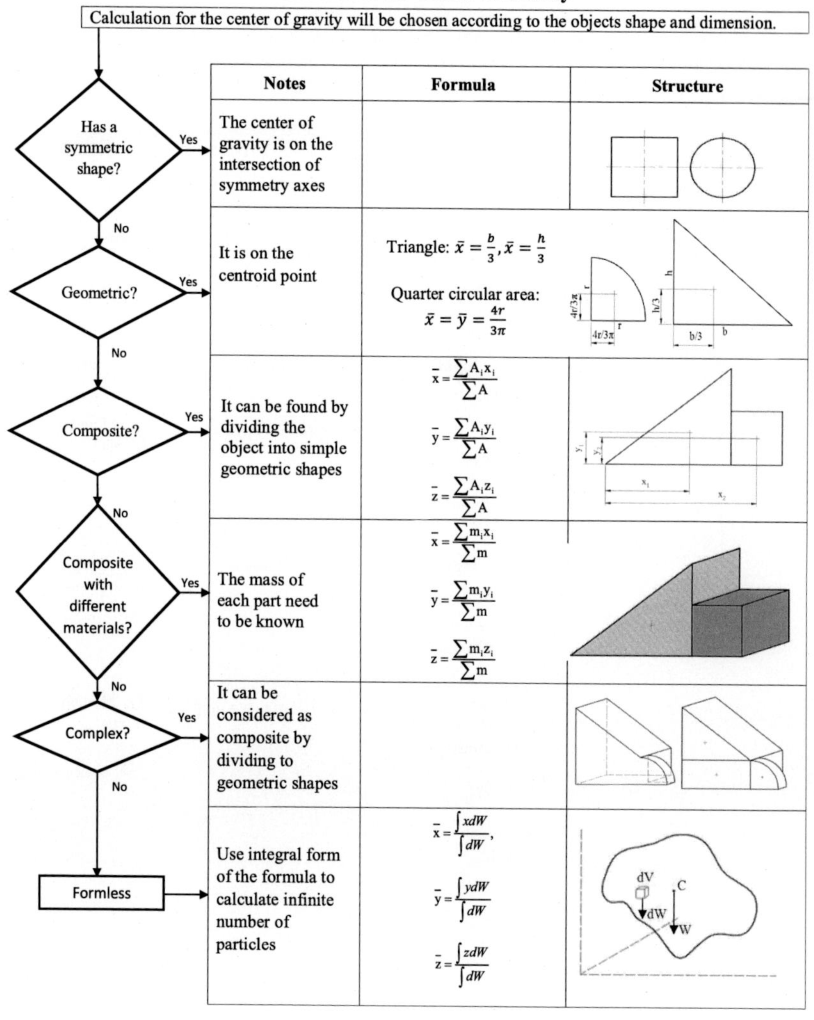

Notes	Formula	Structure
The center of gravity is on the intersection of symmetry axes		
It is on the centroid point	Triangle: $\bar{x} = \frac{b}{3}, \bar{x} = \frac{h}{3}$ Quarter circular area: $\bar{x} = \bar{y} = \frac{4r}{3\pi}$	
It can be found by dividing the object into simple geometric shapes	$\bar{x} = \frac{\sum A_i x_i}{\sum A}$ $\bar{y} = \frac{\sum A_i y_i}{\sum A}$ $\bar{z} = \frac{\sum A_i z_i}{\sum A}$	
The mass of each part need to be known	$\bar{x} = \frac{\sum m_i x_i}{\sum m}$ $\bar{y} = \frac{\sum m_i y_i}{\sum m}$ $\bar{z} = \frac{\sum m_i z_i}{\sum m}$	
It can be considered as composite by dividing to geometric shapes		
Use integral form of the formula to calculate infinite number of particles	$\bar{x} = \frac{\int x dW}{\int dW}$, $\bar{y} = \frac{\int y dW}{\int dW}$ $\bar{z} = \frac{\int z dW}{\int dW}$	

Calculation for Moment of Inertia

The mass moment of inertia is one measure of the distribution of the mass of an object relative to a given axis. The mass moment of inertia is denoted by *I*. Moment of inertia determines the torque needed for a desired angular acceleration about a rotational axis. It depends on the body's mass distribution and the axis chosen, with larger moments requiring more torque to change the body's rotation.

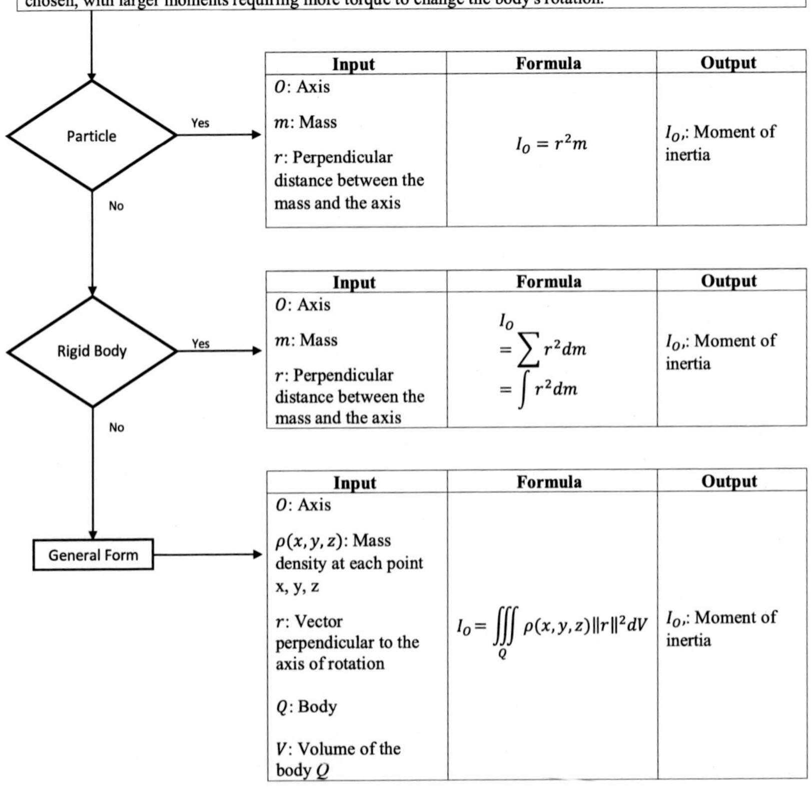

Input	Formula	Output
O: Axis m: Mass r: Perpendicular distance between the mass and the axis	$I_O = r^2 m$	I_O,: Moment of inertia

Input	Formula	Output
O: Axis m: Mass r: Perpendicular distance between the mass and the axis	$I_O = \sum r^2 dm = \int r^2 dm$	I_O,: Moment of inertia

Input	Formula	Output
O: Axis $\rho(x, y, z)$: Mass density at each point x, y, z r: Vector perpendicular to the axis of rotation Q: Body V: Volume of the body Q	$I_O = \iiint_Q \rho(x, y, z) \lVert r \rVert^2 dV$	I_O,: Moment of inertia

Calculation for Flow

Flow type is classified according to Reynolds number. Flow rate can be calculated according to the head drop, pressure drop, and other methods.

Type of the Flow

Terms	Formula(s)	Notes	
• Re = Reynolds number (nondimensional). • ρ = density (kg/m3). • v_s = velocity based on the actual cross-sectional area of the duct or pipe (m/s). • μ = dynamic viscosity (Ns/m^2). • d = characteristic length (m). • v = kinematic viscosity (m2/s).	$Re = \frac{\rho v_s d}{\mu}$ or $Re = \frac{v_s d}{v}$	After calculating the Re for the follow, the follow type can be classified according to the following table.	
		Reynolds number	**Type of flow**
		Re < 2300	Laminar flow
		2300 < Re < 4000	Transition flow
		Re>4000	Turbulent flow

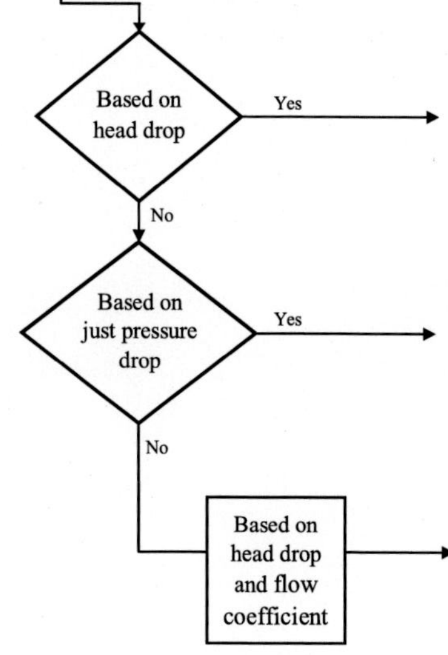

Flow rate

Based on head drop? Yes →

Terms	Formula(s)
• Q: Flow rate (m^3/s) • C: Discharge coefficient • g: Gravity (9.8 m/s^2) • Δh: Head drop (m) • A: Area (m^2) • D: Diameter (m)	$Q = C.\sqrt{2.g.\Delta h}.A$ or $Q = C.\sqrt{2.g.\Delta h}.D^2.\frac{\pi}{4}$
• Q: Flow rate (m^3/s) • C: Discharge coefficient • ΔP: Pressure drop (pa) • d: Density (kg/m^3) • A: Area (m^2) • D: Diameter (m)	$Q = C.\sqrt{2.\frac{\Delta P}{d}}.A$ or $Q = C.\sqrt{2.\frac{\Delta P}{d}}.D^2.\frac{\pi}{4}$
• Q: Flow rate (m^3/s) • K: Flow coefficient (pa$^{-0.5}$) • ΔP: Pressure drop (pa) • SG: Specific gravity	$Q = K.\sqrt{\frac{\Delta P}{SG}}$

No → Based on just pressure drop? Yes → second row of the table above.

No → Based on head drop and flow coefficient → third row of the table above.

Chapter 4

Mechanical Failure Modes

ABSTRACT

Mechanical parts fail depending on the mechanical failure modes. It seems quite straightforward but one should understand that the mechanical parts have a life and this life can be determined by the designer. It is always possible to give an infinite life to a part, but this will make the part big and expensive. In order the optimize the machine, different parts of the machine should have similar lengths of life, or depending on the project some parts on purposely have shorter or longer life spans. If there are two parts working together and if one of them is easier to replace, this part is designed to be knowingly weak. With the same understanding if the machine is very critical, a UAV, the parts are designed not to fail up to a certain time. Mechanical failure modes can be grouped into two categories: (1) instant, force or moment dependent, (2) time dependent. Bending, shear, torsion, and buckling happens instantly if the force/moment applied to the part is more than the design criteria. It is quite easy to calculate these failure modes, since there is no uncertainty. On the other hand, erosion, corrosion, fatigue, creep, and thermal shock are time and environmental dependent and cannot be controlled easily throughout the life of the part, and moreover more than one can occur at the same time causing the life of a part to decrease exponentially.

Depending on the working conditions of the part, the mechanical failure modes should be calculated and the life of the part should be known. For time-dependent failures, it is possible to find the life of a part with experimentation.

Mechatronic Components. https://doi.org/10.1016/B978-0-12-814126-7.00004-9

Bending

Bending is the behavior of a structural element subjected to an external load applied perpendicularly to a longitudinal axis of the element.

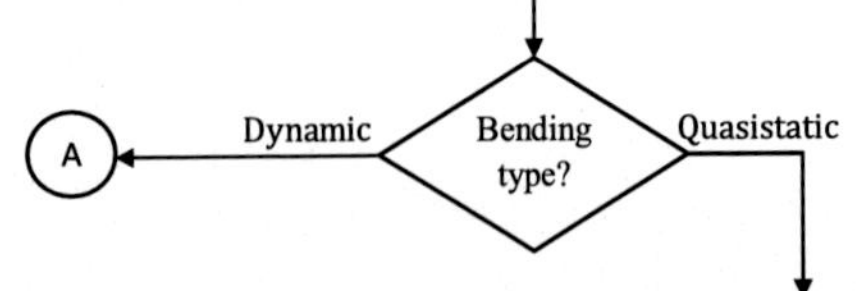

Step 1		
Terms	**Procedure**	**Output**
• E: Young's modulus (N/m²) • I: Area moment of inertia of the cross section (m⁴) • M: Internal bending moment in the beam	$\frac{d^2w(x)}{dx^2} = \frac{M(x)}{E(x)I(x)}$	$\frac{d^2w(x)}{dx^2}$ The second derivative of its deflected shape with respect to x is interpreted as its curvature.

Step 2		
Terms	**Formula(s)**	**Output**
• E: Young's modulus (N/m²) • I: Area moment of inertia of the cross section (m⁴)	$EI\frac{d^4w(x)}{dx^4} = q(x)$	$q(x)$: Applied traverse load (N)

Step 3		
Terms	**Formula(s)**	**Output**
• E: Young's modulus (N/m²) • I: Area moment of inertia of the cross section (m⁴) $\frac{d^2w(x)}{dx^2}$	$M(x) = -EI\frac{d^2w}{dx^2}$ $Q(x) = \frac{dM}{dx}$	M: Bending moment (Nm) Q: Shear force (N)

Step 4		
Input (Unit)	Formula	Output
M: Moment about the neutral axis (Nm) y: Perpendicular distance to the neutral axis (m) I_x: Second moment of area about the neutral axis x	$\sigma = \frac{My}{I_x}$	σ: Bending stress (N/m^2)

A

Step1		
Input (Unit)	Formula	Output
E: Young's modulus (N/m^2) I: Area moment of inertia of the cross section (m^4) w(x,t): Deflection of the neutral axis of the beam m: Mass per unit lenght of the beam (m)	$EI\frac{\partial^4 w}{\partial x^4} + m\frac{\partial^2 w}{\partial t^2} = q(x,t)$	$q(x,t)$: Applied traverse load (N)

Step 2		
Input (Unit)	Formula	Output
E: Young's modulus I: Area moment of inertia of the cross section (m^4)	$EI\frac{\partial^4 w}{\partial x^4} + m\frac{\partial^2 w}{\partial t^2} = 0$	$q(x)$: Applied traverse load (for free vibrations)

Step 3		
Input (Unit)	Formula	Output
w(x,t): Deflection of the neutral axis of the beam m: Mass per unit lenght of the beam (m) ω: Açısal hız (rad/s)	$\frac{\partial^2 w}{\partial t^2} = -\omega^2 w(x,t)$	$\frac{\partial^2 w}{\partial t^2}$ (for free vibrations)

Shear

Shear stress is a stress state where the stress is parallel to the surface of the material as opposed to normal stress when the stress is vertical to the surface.

General Shear Stress		
Terms	**Formula(s)**	**Notes**
• τ: Shear stress (Pa) • F: The force applied (N) • A: The cross-sectional area of material with area parallel to the applied force vector (m^2)	$\tau = \frac{F}{A}$	It is known as average shear stress.

Pure Shear Stress		
Step 1		
Terms	**Formula(s)**	**Notes**
• G :Shear modulus (Pa) • E :Young's modulus (Pa) • v :Poisson's ratio	$G = \frac{E}{2(1+v)}$	It is related to pure shear strain.

↓

Step 2		
Terms	**Formula(s)**	**Notes**
• τ: Shear stress (Pa) • Γ: Shear strain • G: Shear modulus (Pa)	$\tau = \gamma G$	It is related to pure shear strain.

Beam Shear Stress		
Terms	**Formula(s)**	**Notes**
• τ: Shear stress (Pa) • V: Total shear force at the location in question (N) • Q: Statical moment of area (m^3) • t: Thickness in the material perpendicular to the shear (m) • I = Moment of inertia of the entire cross-sectional area (m^4)	$\tau = \frac{VQ}{It}$	It is defined as the internal shear stress of a beam caused by the shear force applied to the beam.

Torsion

Torsion is the twisting of an object due to an applied torque and it is expressed in newton meters (Nm). Torque is constant and transmitted along the object by each section trying to shear over its neighbor.

Step 1

Terms	Procedure	Notes	
		Material	S_{syp}
• S_{yp}: Tensile yield Strength (Pa) • S_{syp}: Shear yield point (Pa)	1. Find the S_{yp} of the material from the tables. 2. Use the tensile-shear yield tables (like the one shown) to find S_{syp}.	Wrought Steel and alloy steel	$0.58 \times S_{yp}$
		Ductile Iron	$0.75 \times S_{yp}$
		Aluminum/ and alloys	$0.55 \times S_{yp}$

Step 2

Terms	Formula(s)	
• J: Polar moment of inertia (m^4).	• Solid shaft of radius R, diameter D.	$J = \frac{\pi R^2}{2} = \frac{\pi D^2}{32}$ 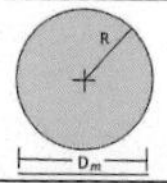
	• Hollow shaft with inner radius R_i and outer radius R_o.	$J = \frac{\pi(R_0^4 - R_i^4)}{2} = \frac{\pi(D_0^4 - D_i^4)}{32}$ 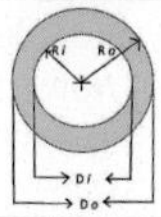
	• Thin-walled tube with t < R/10.	$J = 2\pi R_m^3 t$ 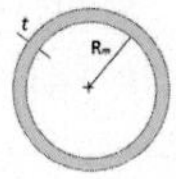

Step 3

Terms	Formula(s)	Notes
• T: Torque (Nm) • τ_{max}: Shear stress (Pa) • J: Polar moment of inertia (m^4) • R: Radius (m)	$T \leq \frac{\tau_{max} J}{R}$	• Ssyp is the point that torsion—caused by the torque—starts to cause plastic deformation, τmax is equal to Ssyp. • The torque must be less than the formula shown. • If the torque is fixed the radius or the profile of the bar can be manipulated.

Buckling

Buckling is a mathematical instability, leading to a failure mode. When an excessive compression force is applied, the part can bend caused by the buckling effect.

Step 1

Terms	**Formula(s)**	**Notes**
• L/r: Slenderness ratio • L: Length of the component (mm) • r: Radius of gyration (mm)	• Circle of radius R: $r = \frac{R}{2}$ • Rectangle of large length R and small length b: $r_{max} = 0.29 \times R$ $r_{min} = 0{,}29$	

Step 2

Terms	**Formula(s)**	**Notes**
• F_{Crit}: Critical buckling Force (N) • L/r: Slenderness ratio • k: Column effective length factor, whose value depends on the conditions of the end support of the column • E: Modulus of elasticity (N/mm^2). • I: Area moment of inertia (depends) • A: Area (mm^2)	$F_{crit} = \frac{k\pi^2 EI}{L^2} = \frac{k\pi^2 EA}{(L/r)^2}$	• Choose k from Table 1. • Table 2 shows the module of elasticity for some materials.

Table 1: Cases of End Restraints and the Associated k Value

Case	1	2	3	4	5
Constraints					
k	4	1	.25	2.046	1

Table 2: Modulus of Elasticity

Material	E
Steel	2.1×10^5
Cast iron	0.8×10^5
Copper	1.2×10^5
Brass	0.9×10^5
Aluminum	0.7×10^5

Step 3		
Terms	**Formula(s)**	**Notes**
• σ_{Crit}: Critical Euler buckling stress (Pa) • F_{Crit}: Critical buckling force (N) • A: Area (mm^2)	$\sigma_{crit} = \frac{F_{crit}}{A}$	

↓

Step 4		
Terms	**Formula(s)**	**Notes**
• λ: Buckling load factor • σ_{Crit}: Critical Euler buckling stress (Pa) • σ_{Curr}: Current stress (Pa)	$\lambda = \frac{\sigma_{crit}}{\sigma_{curr}}$	Check λ from Table 3.

Table 3: Buckling Load Factor (BLF)

BLF Value	Buckling Status	Remarks
$\lambda > 1$	Buckling not predicted	The applied loads are less than the estimated critical loads.
$\lambda = 1$	Buckling predicted	The applied loads are exactly equal to the critical loads. Buckling is expected.
$\lambda < 1$	Buckling predicted	The applied loads exceed the estimated critical loads. Buckling will occur
$-1 < \lambda < 0$	Buckling possible	Buckling is predicted if you reverse the load directions.
$\lambda = -1$	Buckling possible	Buckling is expected if you reverse the load directions
$\lambda < -1$	Buckling not predicted	The applied loads are less than the estimated critical loads, even if you reverse their directions.

↓

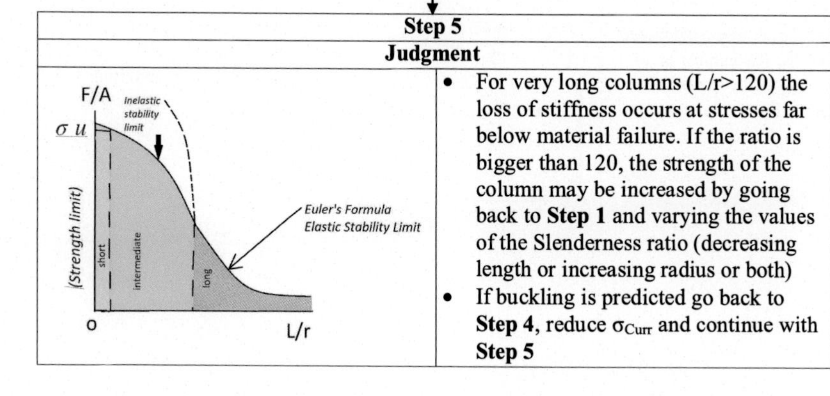

Step 5	
Judgment	
	• For very long columns (L/r>120) the loss of stiffness occurs at stresses far below material failure. If the ratio is bigger than 120, the strength of the column may be increased by going back to **Step 1** and varying the values of the Slenderness ratio (decreasing length or increasing radius or both) • If buckling is predicted go back to **Step 4**, reduce σ_{Curr} and continue with **Step 5**

Calculation for Fatigue

There are many approaches related to fatigue such that stress life and strain life approaches. The stress life approach is applicable for situations involving primarily elastic deformation. Under these conditions the component is expected to have a long lifetime. For situations involving high stresses, high temperatures, or stress concentrations such as notches, where significant plasticity can be involved, and this approach is not appropriate. Rather than stress life approach, the strain life approach is used. As a general rule, consider both elastic and plastic deformation together.

$$\frac{\Delta\epsilon}{2} = \frac{\Delta\epsilon_e}{2} + \frac{\Delta\epsilon_p}{2}$$

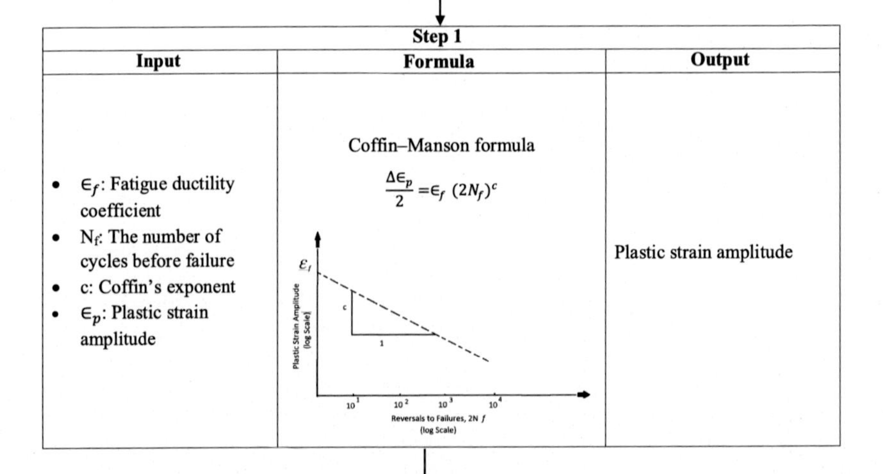

Step 1

Input	Formula	Output
• ϵ_f: Fatigue ductility coefficient • N_f: The number of cycles before failure • c: Coffin's exponent • ϵ_p: Plastic strain amplitude	Coffin–Manson formula $\frac{\Delta\epsilon_p}{2} = \epsilon_f\,(2N_f)^c$	Plastic strain amplitude

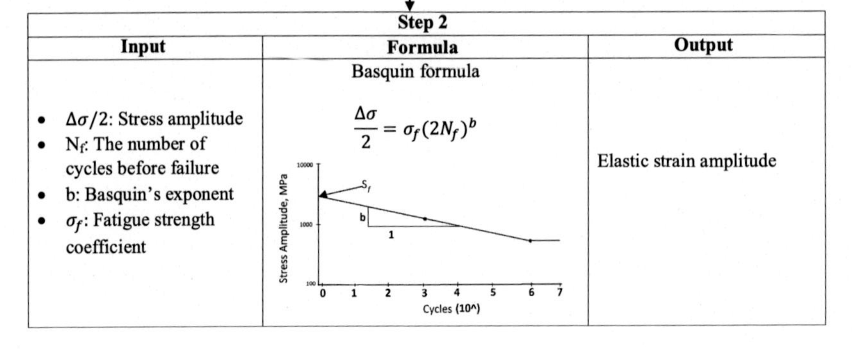

Step 2

Input	Formula	Output
• $\Delta\sigma/2$: Stress amplitude • N_f: The number of cycles before failure • b: Basquin's exponent • σ_f: Fatigue strength coefficient	Basquin formula $\frac{\Delta\sigma}{2} = \sigma_f(2N_f)^b$	Elastic strain amplitude

Step 3		
Input	**Formula**	**Output**
• E: Young modulus • σ_a: Stress amplitude • $\Delta\sigma$: Stress range	1-D elastic loading $\frac{\Delta\epsilon_e}{2} = \frac{\Delta\sigma}{2E} = \frac{\sigma_a}{E}$ Apply this to Basquin law and combine elastic and plastic parts.	General expression containing elastic and plastic parts: $\frac{\Delta\epsilon}{2} = \frac{\sigma_f}{E}(2N_f)^b + \epsilon_f\,(2N_f)^c$

Step 4		
Input	**Formula**	**Output**
• E: Young modulus • ϵ_f: Fatigue ductility coefficient • σ_f: Fatigue strength coefficient	The transition life is found by equalizing the plastic and elastic strain: $2N_t = \left(\frac{\epsilon_f\, E}{\sigma_f}\right)^{1/(b-c)}$ N>>N_t : long component lifetimes N<<N_t : short component lifetimes Strain Amplitude (log scale) $\Delta\varepsilon/2$ $2N_t$ (log scale)	N_t

Step 5		
Input	**Formula**	**Output**
• Δa: Amount of crack growth • ΔK: K-range nominal stress intensity range • C,m: Experimentally determined material parameter	Paris law $\frac{\Delta a}{\Delta N} \rightarrow \frac{da}{dN} = C(\Delta K)^m$	Paris regime can be carried out to determine the number of cycles to failure.

Step 6		
Input	**Formula**	**Output**
• Y: Geometrical parameter	$\Delta K = Y\Delta\sigma\sqrt{\pi\alpha}$	K: Stress intensity factor

↓

Step 7		
Input	**Formula**	**Output**
• K: Stress intensity factor • Paris law Combining the two laws and then integrating the equation	$N_f = \frac{1}{CY^m(\Delta\sigma)^2\pi} ln \frac{a_f}{a_0}$	N_f: Remaining number of cycles to fracture

Creep

Creep is the tendency of a solid material to move slowly or deform permanently under the influence of mechanical stresses and high temperature.

Step 1		
Input	**Formula**	**Output**
• R : Universal gas constant • N_A: Avogadro constant • R= 8.31 x 103 (N*m)/(kmol * K) • $N_A = 6.02 \times 1023$.	$k = \frac{R}{N_A}$	k = Boltzmann constant

Step 2		
Input	**Formula**	**Output**
• T: Absolute temperature(K) • ξ˙: Strain rate By plotting natural log of strain rate against the reciprocal of temperature, activation energy will be obtained.	From the figure below we can easily find the output **Note:** This data obtained experimentally 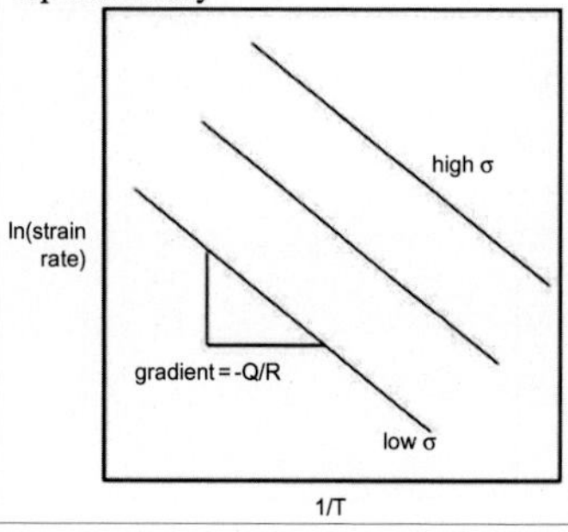 high σ ln(strain rate) gradient = -Q/R low σ 1/T	Q : Activation energy

Step 3		
Input	**Formula**	**Output**
• σ : Applied stress (N/m^2) • $\dot{\xi}$: Strain rate Stress exponent can be obtained by plotting the strain rate as a function of stress.	From the figure below the output is obtained. **Note:** This data obtained experimentally. 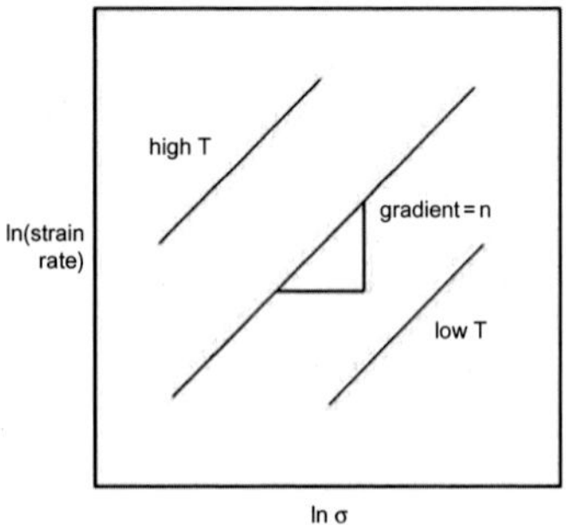 high T gradient = n ln(strain rate) low T ln σ	n: Activation energy

Step 4		
Input	**Formula**	**Output**
• A: Constant dependent on the creep mechanism and material • And all other findings ($k,T,\dot{\xi},\sigma,n,Q$)	$\dot{\varepsilon} = A\,\sigma^{n}\,e^{-\frac{Q}{kT}}$	$\dot{\varepsilon}$: Creep rate is an irreversible process in which the long molecular chains slide along each other. If anything goes wrong re-do the experiment and check the value of constant A.

Thermal Shock

Thermal shock occurs when a thermal gradient causes different parts of an object to expand by different amounts. This different amount of expansion causes stress inside the part. At some point, this stress can exceed the strength of the material, causing a crack to form.

↓

Step 1		
Input	**Formula(s)**	**Output**
• α: Linear coefficient of thermal expansion (K^{-1}) • T_1: Heating temperature (k) • T_2: Starting temperature (k) • L: Length of specimen	$\varepsilon_{th} = \alpha * (T_2 - T_1) = \frac{\Delta l}{l}$	• ε: Change in length with temperature for a solid material.

↓

Step 2		
Input	**Formula(s)**	**Output**
• ε: Change in length with temperature for a solid material • V: Poisson´s ratio (metal: 0.3–0.4, nonmetal: 0.4–0.5). • E: Young´s modulus of elasticity (N/m^2)	$\sigma_{th} = \frac{\varepsilon_{th} * E}{1 - V}$	σ_{th} :Thermal stress (Pa)

↓

Step 3		
Input	**Formula(s)**	**Output**
• σ_{th} : Thermal stress (Pa) • σ_f : Maximum tension = typical material value (Pa)	$R_f \cong \frac{\sigma_f * k * (1 - V)}{\alpha * E}$ $R_{th} \cong \frac{\sigma_{th} * k * (1 - V)}{\alpha * E}$	• R_{th}: Thermal shock parameter for current thermal stress value (K) • R_f : Thermal shock parameter for critical thermal stress value (K)

↓

Judgment
If $\sigma_{th} > \sigma_f$, R is not met and the material will suffer from a plastic deformation or destruction in cause of a thermal shock, so vary the parameters of ∝, V, and E to get a lower R.

Chapter 5

Materials Properties

ABSTRACT

The mechanical parts can be manufactured from different types of materials. There are basic materials, such as aluminum and copper and there are alloys. It is very important to choose the correct material to manufacture the part since the mechanical properties such as yield strength, elasticity module, and hardness determine the resistance to different failure modes. Moreover, the material properties of the part can also have an effect on the dimensions: if a part needs to be lighter but needs to carry some certain design force, then a higher strength-to-weight ratio material should be used.

While it is easier to choose material for force/moment dependent failure modes, it is harder to choose materials for time-dependent failure modes. It should be also noted that the ease of manufacturing must be considered during the material selection process as well as its availability

Mechatronic Components. https://doi.org/10.1016/B978-0-12-814126-7.00005-0

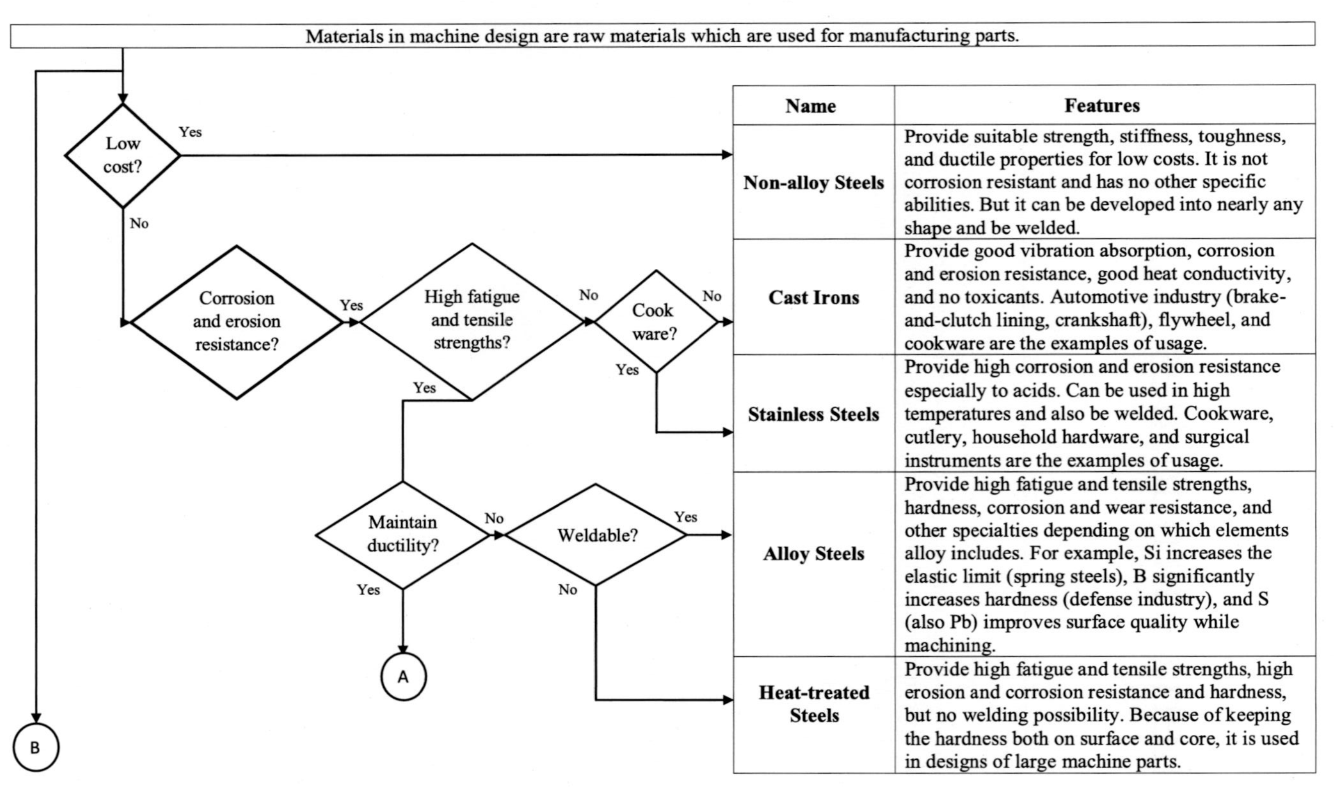

Name	Features
Non-alloy Steels	Provide suitable strength, stiffness, toughness, and ductile properties for low costs. It is not corrosion resistant and has no other specific abilities. But it can be developed into nearly any shape and be welded.
Cast Irons	Provide good vibration absorption, corrosion and erosion resistance, good heat conductivity, and no toxicants. Automotive industry (brake-and-clutch lining, crankshaft), flywheel, and cookware are the examples of usage.
Stainless Steels	Provide high corrosion and erosion resistance especially to acids. Can be used in high temperatures and also be welded. Cookware, cutlery, household hardware, and surgical instruments are the examples of usage.
Alloy Steels	Provide high fatigue and tensile strengths, hardness, corrosion and wear resistance, and other specialties depending on which elements alloy includes. For example, Si increases the elastic limit (spring steels), B significantly increases hardness (defense industry), and S (also Pb) improves surface quality while machining.
Heat-treated Steels	Provide high fatigue and tensile strengths, high erosion and corrosion resistance and hardness, but no welding possibility. Because of keeping the hardness both on surface and core, it is used in designs of large machine parts.

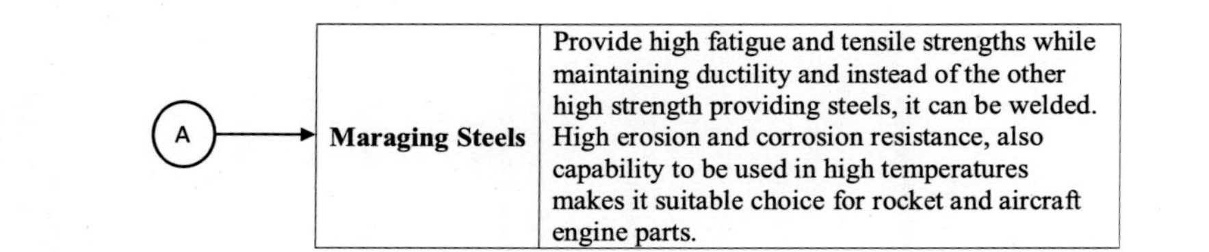
A
Maraging Steels
Provide high fatigue and tensile strengths while maintaining ductility and instead of the other high strength providing steels, it can be welded. High erosion and corrosion resistance, also capability to be used in high temperatures makes it suitable choice for rocket and aircraft engine parts.

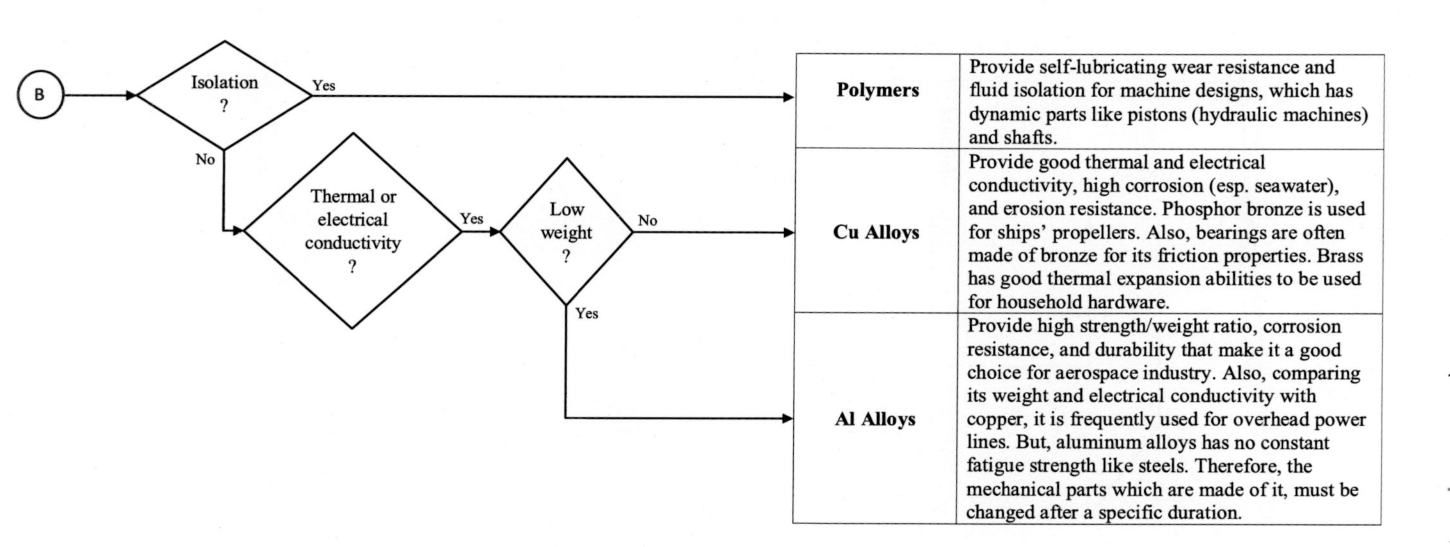
B
Isolation ?
Yes
No
Thermal or electrical conductivity ?
Yes
Low weight ?
No
Yes
Polymers
Provide self-lubricating wear resistance and fluid isolation for machine designs, which has dynamic parts like pistons (hydraulic machines) and shafts.
Cu Alloys
Provide good thermal and electrical conductivity, high corrosion (esp. seawater), and erosion resistance. Phosphor bronze is used for ships' propellers. Also, bearings are often made of bronze for its friction properties. Brass has good thermal expansion abilities to be used for household hardware.
Al Alloys
Provide high strength/weight ratio, corrosion resistance, and durability that make it a good choice for aerospace industry. Also, comparing its weight and electrical conductivity with copper, it is frequently used for overhead power lines. But, aluminum alloys has no constant fatigue strength like steels. Therefore, the mechanical parts which are made of it, must be changed after a specific duration.

Aluminum Types

Aluminum alloys which are in aluminum (Al) are the predominant metal. They are mostly preferred for their high strength-to-weight ratio.

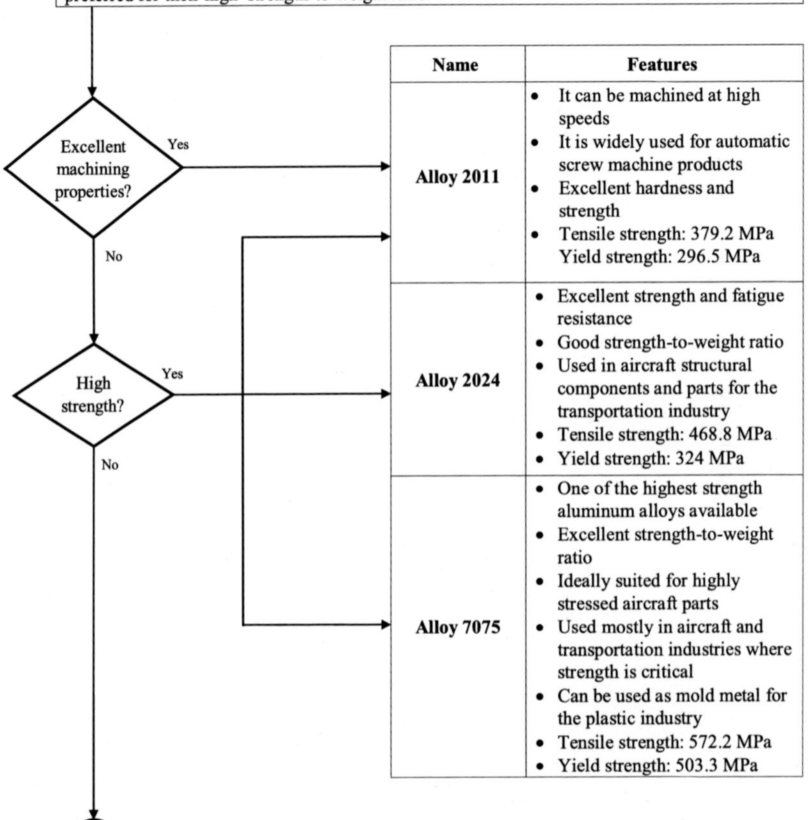

Name	Features
Alloy 2011	• It can be machined at high speeds • It is widely used for automatic screw machine products • Excellent hardness and strength • Tensile strength: 379.2 MPa Yield strength: 296.5 MPa
Alloy 2024	• Excellent strength and fatigue resistance • Good strength-to-weight ratio • Used in aircraft structural components and parts for the transportation industry • Tensile strength: 468.8 MPa • Yield strength: 324 MPa
Alloy 7075	• One of the highest strength aluminum alloys available • Excellent strength-to-weight ratio • Ideally suited for highly stressed aircraft parts • Used mostly in aircraft and transportation industries where strength is critical • Can be used as mold metal for the plastic industry • Tensile strength: 572.2 MPa • Yield strength: 503.3 MPa

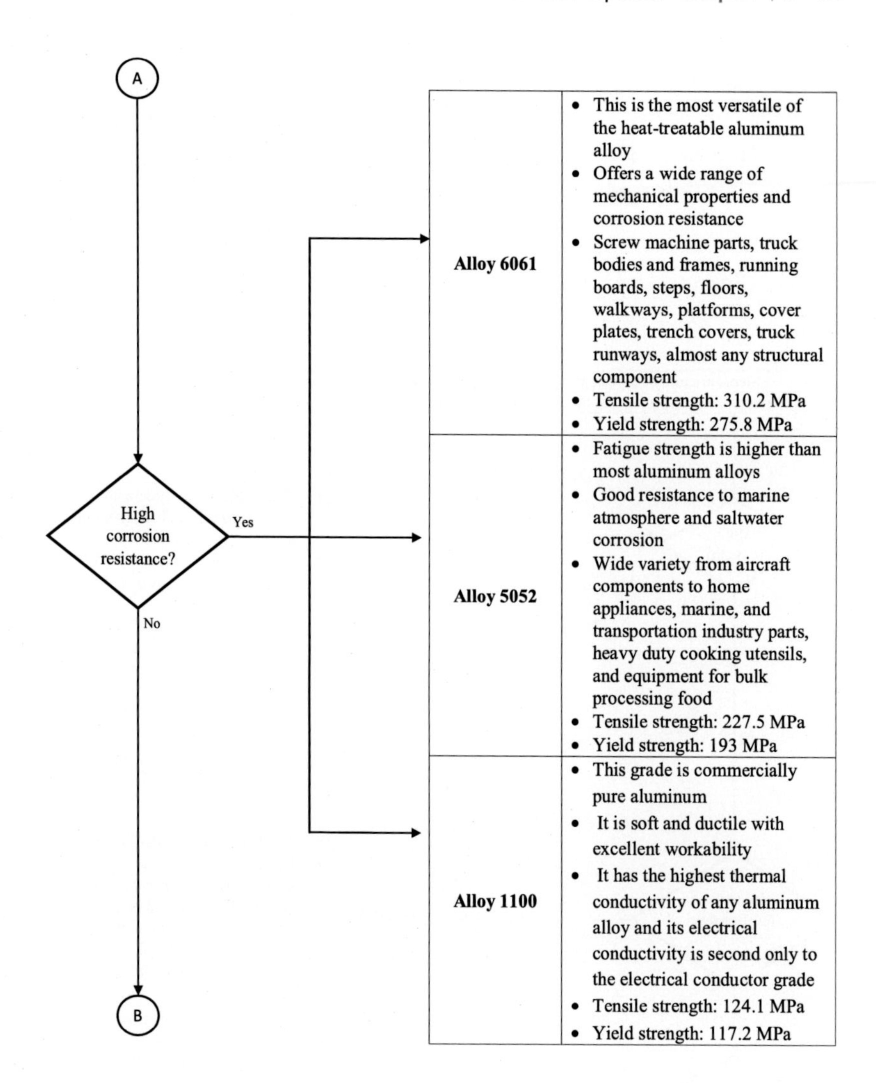
A
High corrosion resistance?
Yes
No
B
Alloy 6061
• This is the most versatile of the heat-treatable aluminum alloy
• Offers a wide range of mechanical properties and corrosion resistance
• Screw machine parts, truck bodies and frames, running boards, steps, floors, walkways, platforms, cover plates, trench covers, truck runways, almost any structural component
• Tensile strength: 310.2 MPa
• Yield strength: 275.8 MPa
Alloy 5052
• Fatigue strength is higher than most aluminum alloys
• Good resistance to marine atmosphere and saltwater corrosion
• Wide variety from aircraft components to home appliances, marine, and transportation industry parts, heavy duty cooking utensils, and equipment for bulk processing food
• Tensile strength: 227.5 MPa
• Yield strength: 193 MPa
Alloy 1100
• This grade is commercially pure aluminum
• It is soft and ductile with excellent workability
• It has the highest thermal conductivity of any aluminum alloy and its electrical conductivity is second only to the electrical conductor grade
• Tensile strength: 124.1 MPa
• Yield strength: 117.2 MPa

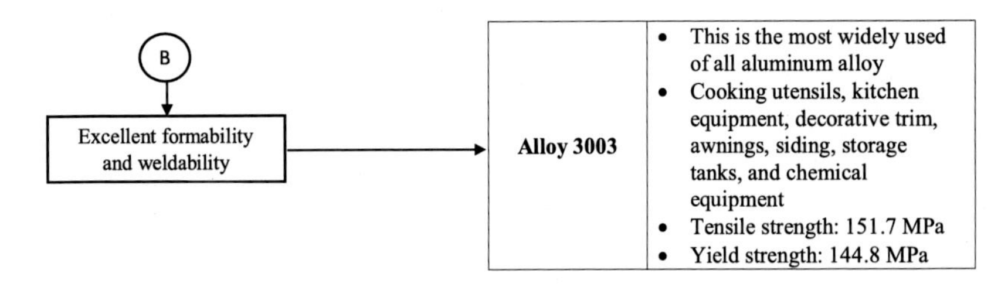
B
Excellent formability and weldability
Alloy 3003
This is the most widely used of all aluminum alloy
Cooking utensils, kitchen equipment, decorative trim, awnings, siding, storage tanks, and chemical equipment
Tensile strength: 151.7 MPa
Yield strength: 144.8 MPa

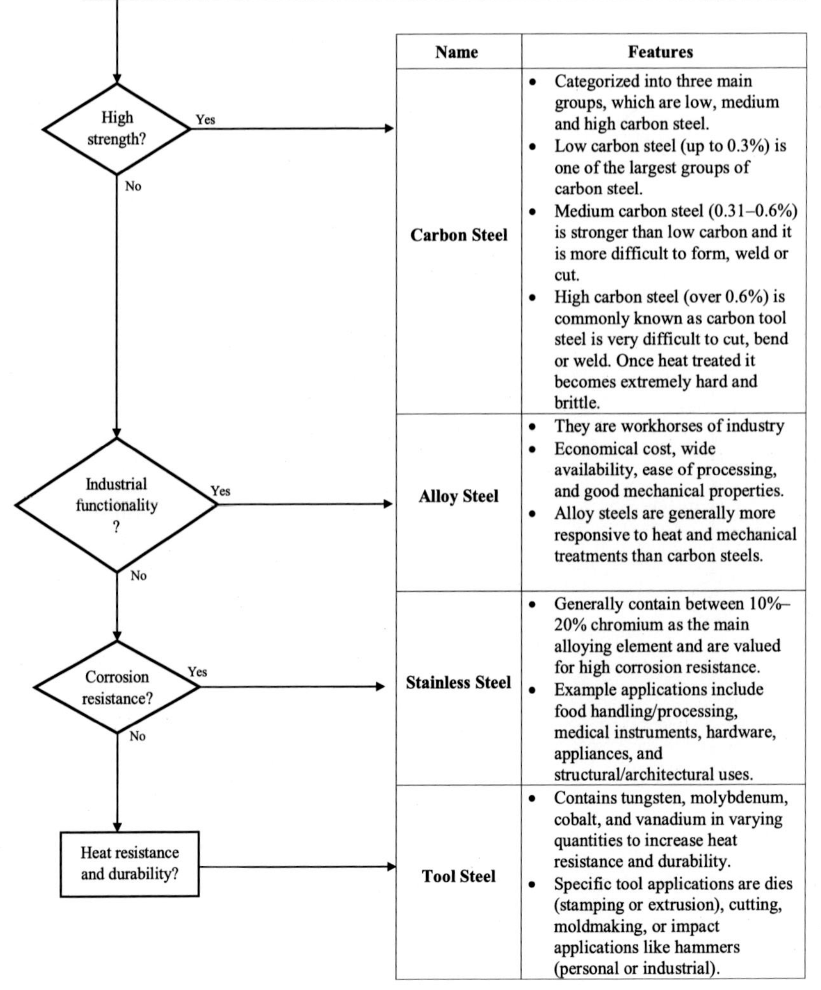
Steel Types
Steels are alloys of iron and other elements, primarily carbon, widely used in construction and other applications because of their high tensile strength and low cost.
High strength?
Yes
No
Industrial functionality ?
Yes
No
Corrosion resistance?
Yes
No
Heat resistance and durability?
Name
Features
Carbon Steel
Categorized into three main groups, which are low, medium and high carbon steel.
Low carbon steel (up to 0.3%) is one of the largest groups of carbon steel.
Medium carbon steel (0.31–0.6%) is stronger than low carbon and it is more difficult to form, weld or cut.
High carbon steel (over 0.6%) is commonly known as carbon tool steel is very difficult to cut, bend or weld. Once heat treated it becomes extremely hard and brittle.
Alloy Steel
They are workhorses of industry
Economical cost, wide availability, ease of processing, and good mechanical properties.
Alloy steels are generally more responsive to heat and mechanical treatments than carbon steels.
Stainless Steel
Generally contain between 10%–20% chromium as the main alloying element and are valued for high corrosion resistance.
Example applications include food handling/processing, medical instruments, hardware, appliances, and structural/architectural uses.
Tool Steel
Contains tungsten, molybdenum, cobalt, and vanadium in varying quantities to increase heat resistance and durability.
Specific tool applications are dies (stamping or extrusion), cutting, moldmaking, or impact applications like hammers (personal or industrial).

Titanium Types

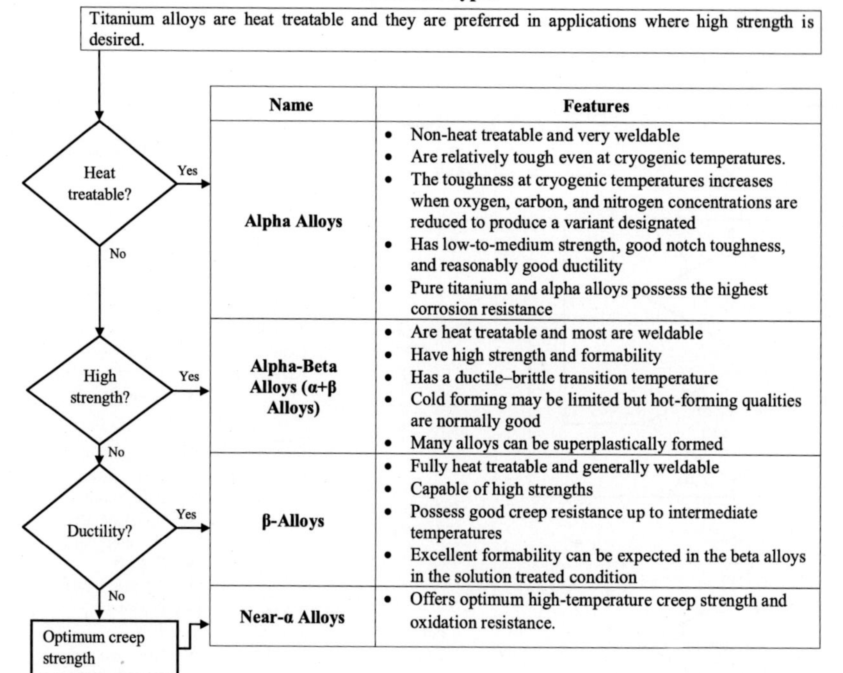

Name	Features
Alpha Alloys	• Non-heat treatable and very weldable • Are relatively tough even at cryogenic temperatures. • The toughness at cryogenic temperatures increases when oxygen, carbon, and nitrogen concentrations are reduced to produce a variant designated • Has low-to-medium strength, good notch toughness, and reasonably good ductility • Pure titanium and alpha alloys possess the highest corrosion resistance
Alpha-Beta Alloys (α+β Alloys)	• Are heat treatable and most are weldable • Have high strength and formability • Has a ductile–brittle transition temperature • Cold forming may be limited but hot-forming qualities are normally good • Many alloys can be superplastically formed
β-Alloys	• Fully heat treatable and generally weldable • Capable of high strengths • Possess good creep resistance up to intermediate temperatures • Excellent formability can be expected in the beta alloys in the solution treated condition
Near-α Alloys	• Offers optimum high-temperature creep strength and oxidation resistance.

Plastic Types and Features

Plastic types are classified based of the recyclability, resistance to impact and oxygen barrier. Some plastic types are a source of risk over the health, this factor too is used in classification of the plastic types. Resin identification codes were added under the names.

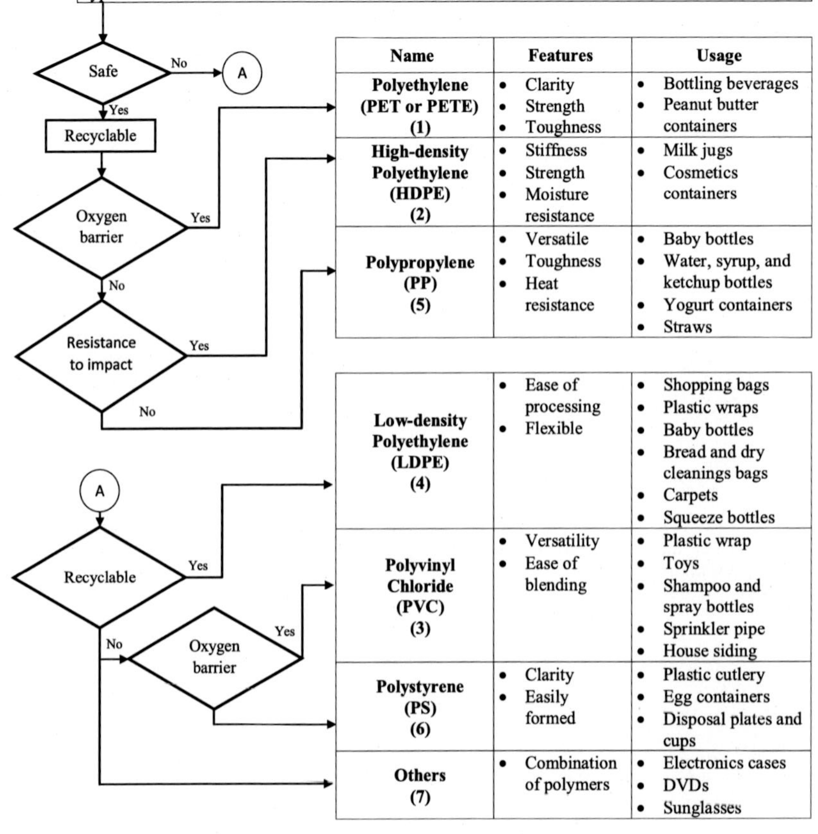

Name	Features	Usage
Polyethylene (PET or PETE) (1)	• Clarity • Strength • Toughness	• Bottling beverages • Peanut butter containers
High-density Polyethylene (HDPE) (2)	• Stiffness • Strength • Moisture resistance	• Milk jugs • Cosmetics containers
Polypropylene (PP) (5)	• Versatile • Toughness • Heat resistance	• Baby bottles • Water, syrup, and ketchup bottles • Yogurt containers • Straws
Low-density Polyethylene (LDPE) (4)	• Ease of processing • Flexible	• Shopping bags • Plastic wraps • Baby bottles • Bread and dry cleanings bags • Carpets • Squeeze bottles
Polyvinyl Chloride (PVC) (3)	• Versatility • Ease of blending	• Plastic wrap • Toys • Shampoo and spray bottles • Sprinkler pipe • House siding
Polystyrene (PS) (6)	• Clarity • Easily formed	• Plastic cutlery • Egg containers • Disposal plates and cups
Others (7)	• Combination of polymers	• Electronics cases • DVDs • Sunglasses

3D Filament Types

3D printer filaments are raw material for 3D printing. Their durability, friction coefficient, flexibility, and other properties define their usage.

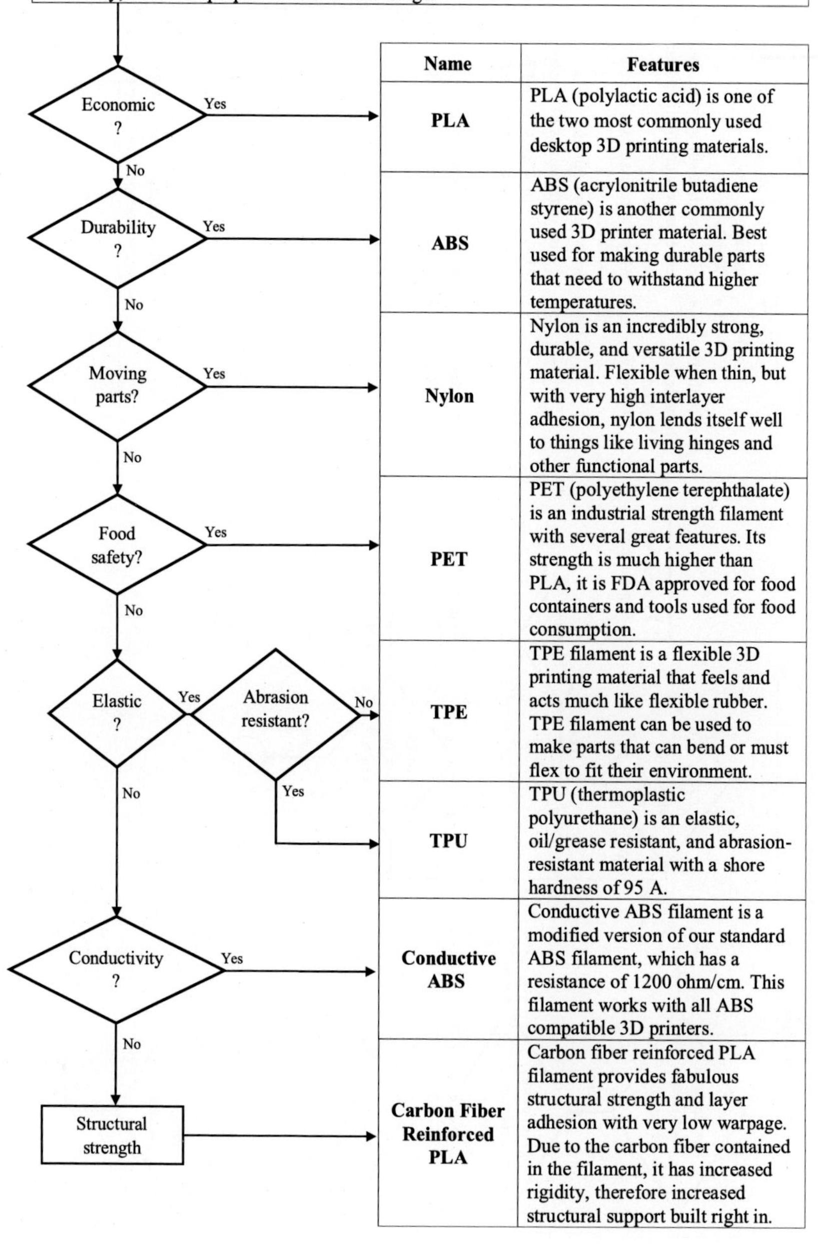

Name	Features
PLA	PLA (polylactic acid) is one of the two most commonly used desktop 3D printing materials.
ABS	ABS (acrylonitrile butadiene styrene) is another commonly used 3D printer material. Best used for making durable parts that need to withstand higher temperatures.
Nylon	Nylon is an incredibly strong, durable, and versatile 3D printing material. Flexible when thin, but with very high interlayer adhesion, nylon lends itself well to things like living hinges and other functional parts.
PET	PET (polyethylene terephthalate) is an industrial strength filament with several great features. Its strength is much higher than PLA, it is FDA approved for food containers and tools used for food consumption.
TPE	TPE filament is a flexible 3D printing material that feels and acts much like flexible rubber. TPE filament can be used to make parts that can bend or must flex to fit their environment.
TPU	TPU (thermoplastic polyurethane) is an elastic, oil/grease resistant, and abrasion-resistant material with a shore hardness of 95 A.
Conductive ABS	Conductive ABS filament is a modified version of our standard ABS filament, which has a resistance of 1200 ohm/cm. This filament works with all ABS compatible 3D printers.
Carbon Fiber Reinforced PLA	Carbon fiber reinforced PLA filament provides fabulous structural strength and layer adhesion with very low warpage. Due to the carbon fiber contained in the filament, it has increased rigidity, therefore increased structural support built right in.

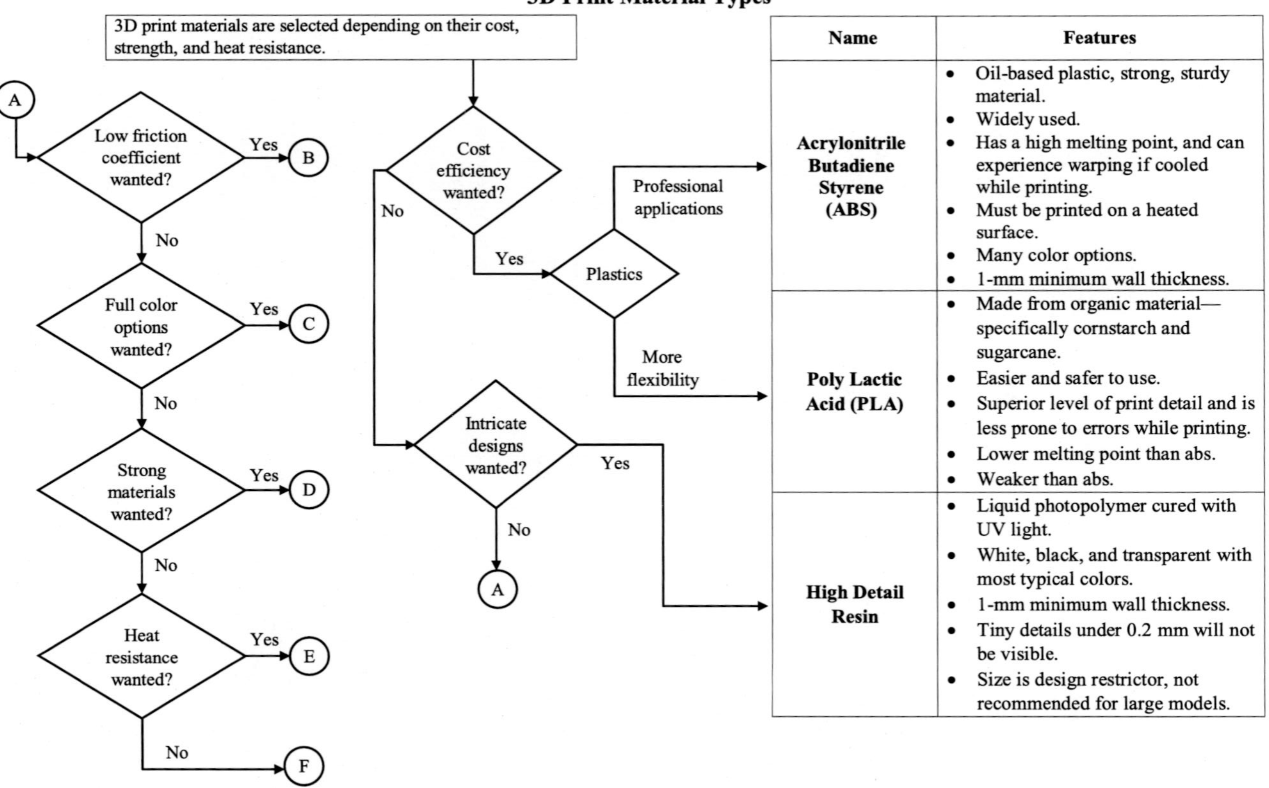

Name	Features
Acrylonitrile Butadiene Styrene (ABS)	• Oil-based plastic, strong, sturdy material. • Widely used. • Has a high melting point, and can experience warping if cooled while printing. • Must be printed on a heated surface. • Many color options. • 1-mm minimum wall thickness.
Poly Lactic Acid (PLA)	• Made from organic material—specifically cornstarch and sugarcane. • Easier and safer to use. • Superior level of print detail and is less prone to errors while printing. • Lower melting point than abs. • Weaker than abs.
High Detail Resin	• Liquid photopolymer cured with UV light. • White, black, and transparent with most typical colors. • 1-mm minimum wall thickness. • Tiny details under 0.2 mm will not be visible. • Size is design restrictor, not recommended for large models.

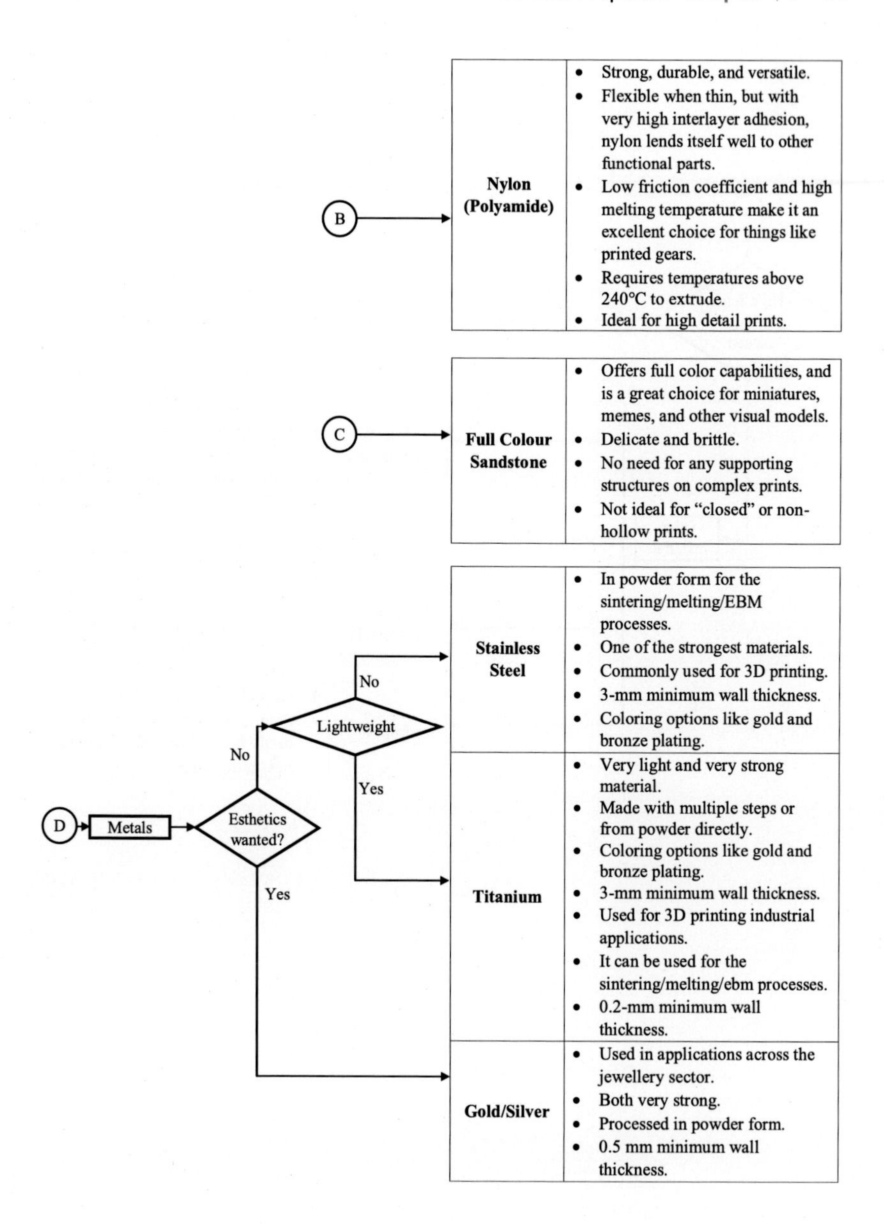
B
Nylon (Polyamide)
Strong, durable, and versatile.
Flexible when thin, but with very high interlayer adhesion, nylon lends itself well to other functional parts.
Low friction coefficient and high melting temperature make it an excellent choice for things like printed gears.
Requires temperatures above 240°C to extrude.
Ideal for high detail prints.
C
Full Colour Sandstone
Offers full color capabilities, and is a great choice for miniatures, memes, and other visual models.
Delicate and brittle.
No need for any supporting structures on complex prints.
Not ideal for “closed” or non-hollow prints.
D
Metals
Esthetics wanted?
No
Yes
Lightweight
No
Yes
Stainless Steel
In powder form for the sintering/melting/EBM processes.
One of the strongest materials.
Commonly used for 3D printing.
3-mm minimum wall thickness.
Coloring options like gold and bronze plating.
Titanium
Very light and very strong material.
Made with multiple steps or from powder directly.
Coloring options like gold and bronze plating.
3-mm minimum wall thickness.
Used for 3D printing industrial applications.
It can be used for the sintering/melting/ebm processes.
0.2-mm minimum wall thickness.
Gold/Silver
Used in applications across the jewellery sector.
Both very strong.
Processed in powder form.
0.5 mm minimum wall thickness.

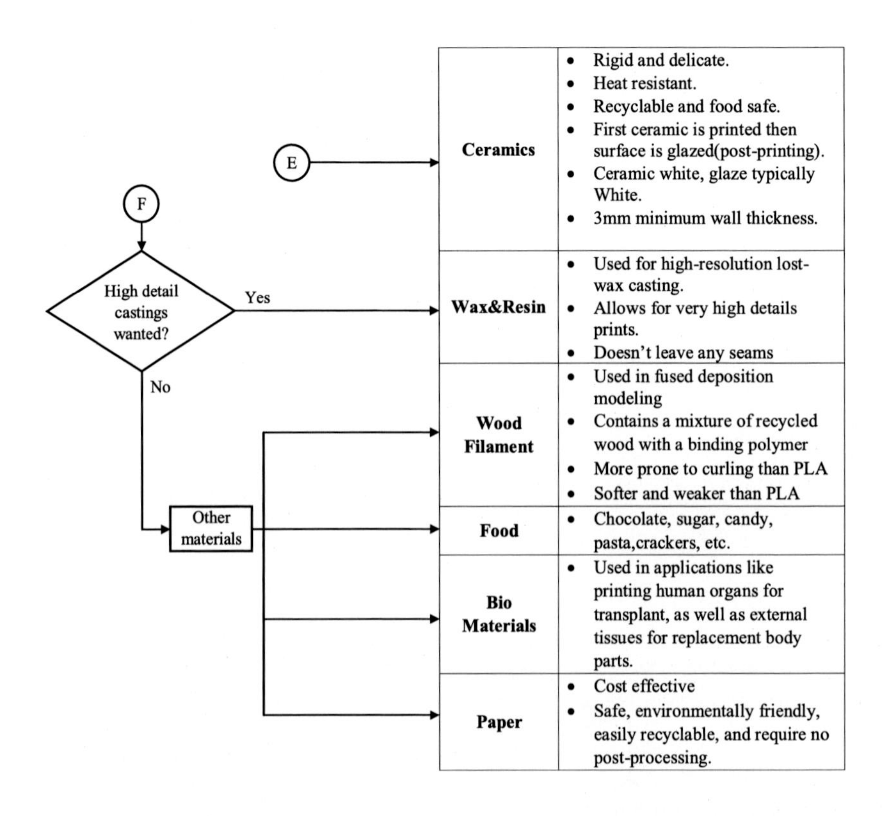

E
F
High detail castings wanted?
Yes
No
Other materials
Ceramics
• Rigid and delicate.
• Heat resistant.
• Recyclable and food safe.
• First ceramic is printed then surface is glazed(post-printing).
• Ceramic white, glaze typically White.
• 3mm minimum wall thickness.
Wax&Resin
• Used for high-resolution lost-wax casting.
• Allows for very high details prints.
• Doesn't leave any seams
Wood Filament
• Used in fused deposition modeling
• Contains a mixture of recycled wood with a binding polymer
• More prone to curling than PLA
• Softer and weaker than PLA
Food
• Chocolate, sugar, candy, pasta,crackers, etc.
Bio Materials
• Used in applications like printing human organs for transplant, as well as external tissues for replacement body parts.
Paper
• Cost effective
• Safe, environmentally friendly, easily recyclable, and require no post-processing.

Chapter 6

Manufacturing Processes

ABSTRACT

A mechanical part can be manufactured in different ways, but even though the geometry and tolerances are the same, the mechanical properties can be different: consider a part manufactured by casting, milling, or forging. Each of the manufacturing methods has advantages and disadvantages. For example, if a part is manufactured by casting it will be more resilient to vibration and impact, but it will have a lower yield strength compared to the same part that is manufactured by milling. Some of the manufacturing methods produce parts that have mechanical properties dependent on the direction of the part, such as rolling and extrusion. Most of the manufacturing processes use a cutting concept from a bigger material, such as lathe, milling, or drilling. Molding uses the concept of the molten material to be poured into a shape. For molded parts, stress distribution inside the part because of different cooling speeds of the different sections still poses a great challenge. Additive manufacturing methods are gaining more popularity and are becoming more common with the help of cheap filament-type 3D printers. This method is especially becoming important in educational environments and probably in the next decade we will see biomaterial 3D printers becoming common and metal 3D printers becoming more accessible and cheaper to use.

Mechatronic Components. https://doi.org/10.1016/B978-0-12-814126-7.00006-2

Lathe

Lathe is a machine tool which rotates the workpiece on its axis to perform various operations such as cutting, sanding, knurling, drilling, deformation, facing, turning, or grooving. Cutting is done with tools that are applied to the workpiece to create an object which has symmetry about an axis of rotation.

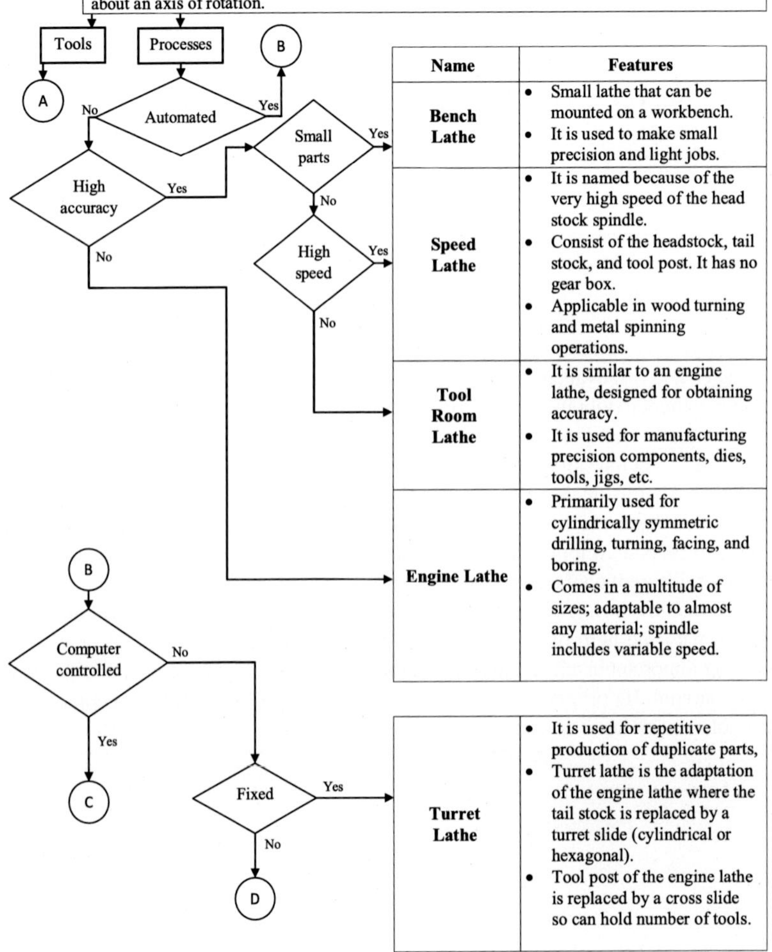

Name	Features
Bench Lathe	• Small lathe that can be mounted on a workbench. • It is used to make small precision and light jobs.
Speed Lathe	• It is named because of the very high speed of the head stock spindle. • Consist of the headstock, tail stock, and tool post. It has no gear box. • Applicable in wood turning and metal spinning operations.
Tool Room Lathe	• It is similar to an engine lathe, designed for obtaining accuracy. • It is used for manufacturing precision components, dies, tools, jigs, etc.
Engine Lathe	• Primarily used for cylindrically symmetric drilling, turning, facing, and boring. • Comes in a multitude of sizes; adaptable to almost any material; spindle includes variable speed.

Turret Lathe	• It is used for repetitive production of duplicate parts, • Turret lathe is the adaptation of the engine lathe where the tail stock is replaced by a turret slide (cylindrical or hexagonal). • Tool post of the engine lathe is replaced by a cross slide so can hold number of tools.

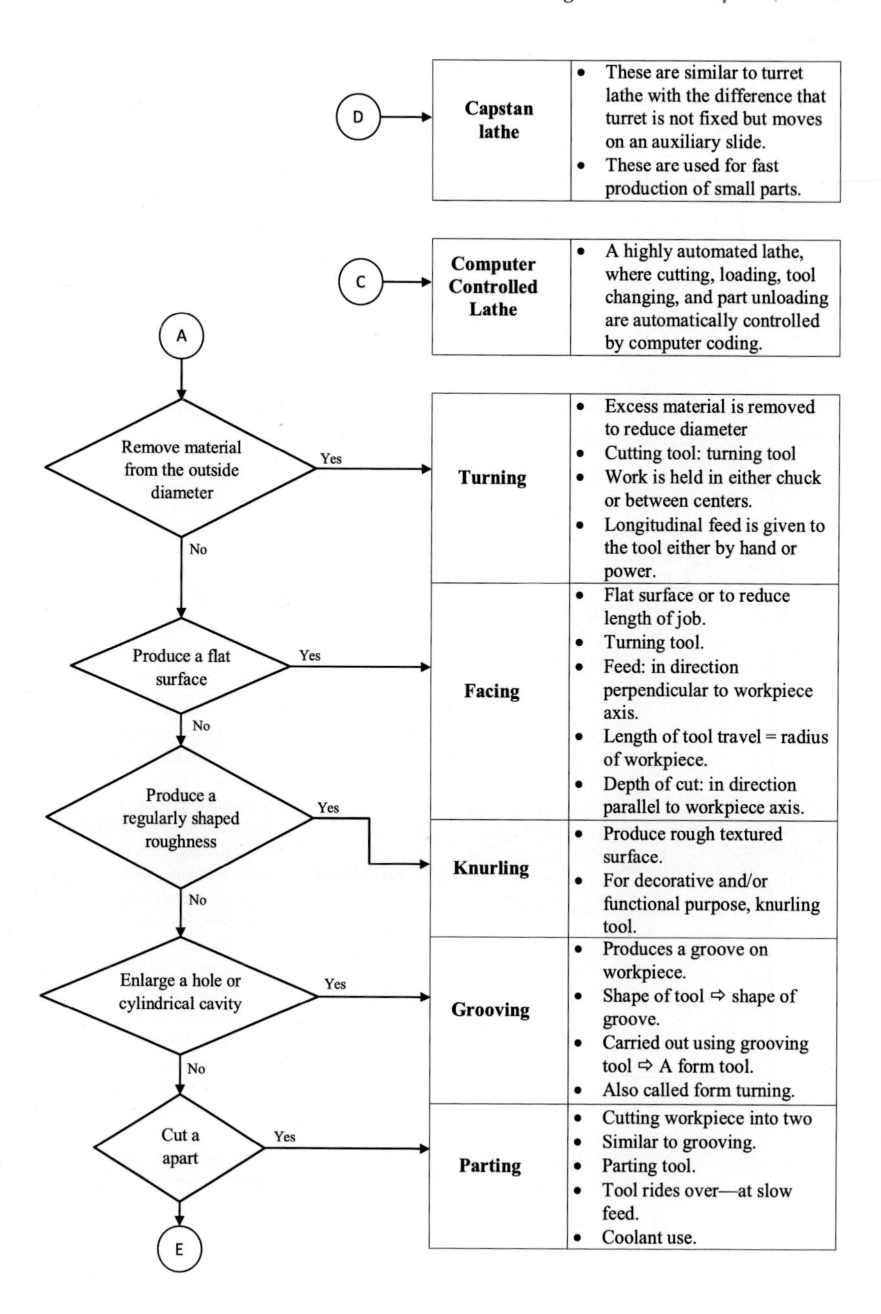
D
Capstan lathe
These are similar to turret lathe with the difference that turret is not fixed but moves on an auxiliary slide.
These are used for fast production of small parts.
C
Computer Controlled Lathe
A highly automated lathe, where cutting, loading, tool changing, and part unloading are automatically controlled by computer coding.
A
Remove material from the outside diameter
Yes
No
Turning
Excess material is removed to reduce diameter
Cutting tool: turning tool
Work is held in either chuck or between centers.
Longitudinal feed is given to the tool either by hand or power.
Produce a flat surface
Yes
No
Facing
Flat surface or to reduce length of job.
Turning tool.
Feed: in direction perpendicular to workpiece axis.
Length of tool travel = radius of workpiece.
Depth of cut: in direction parallel to workpiece axis.
Produce a regularly shaped roughness
Yes
No
Knurling
Produce rough textured surface.
For decorative and/or functional purpose, knurling tool.
Enlarge a hole or cylindrical cavity
Yes
No
Grooving
Produces a groove on workpiece.
Shape of tool ⇨ shape of groove.
Carried out using grooving tool ⇨ A form tool.
Also called form turning.
Cut a apart
Yes
Parting
Cutting workpiece into two
Similar to grooving.
Parting tool.
Tool rides over—at slow feed.
Coolant use.
E

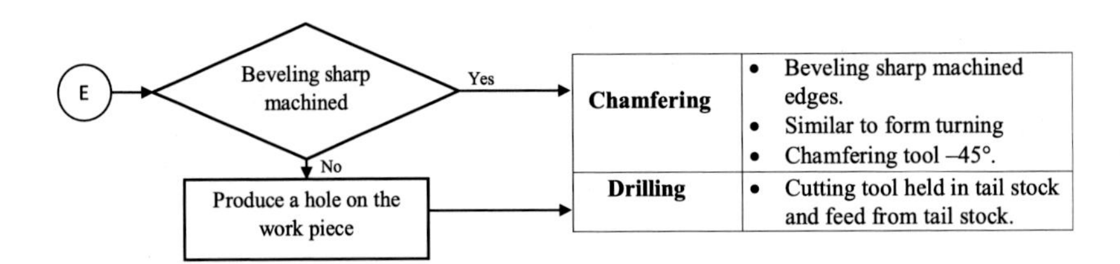

Chamfering	• Beveling sharp machined edges. • Similar to form turning • Chamfering tool –45°.
Drilling	• Cutting tool held in tail stock and feed from tail stock.

Milling

Milling is the machining process of using rotary cutters to remove material from a workpiece advancing (or feeding) in a direction or at an angle with the axis of the tool.

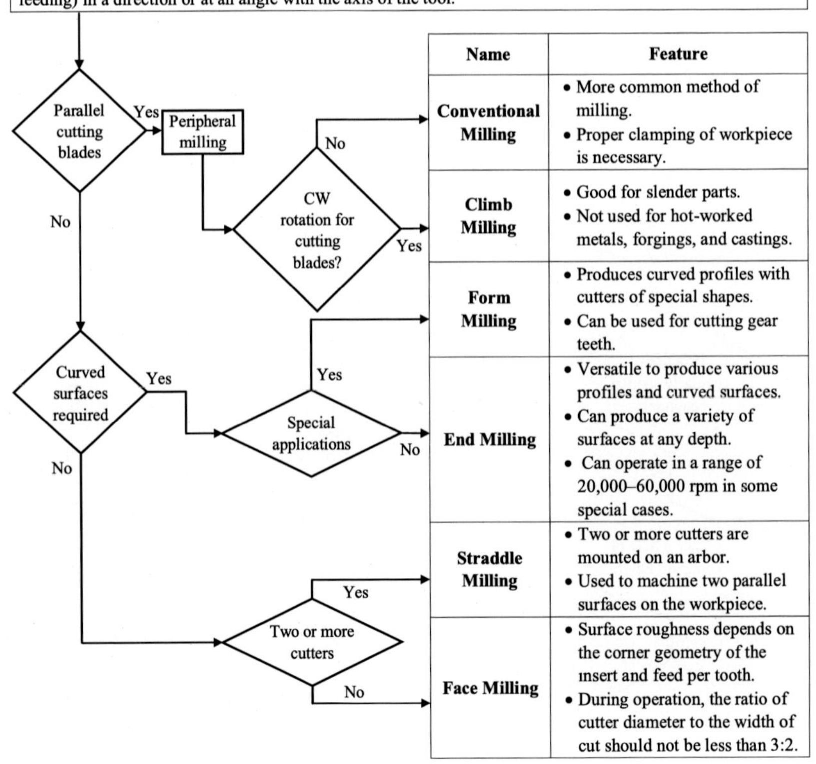

Name	Feature
Conventional Milling	• More common method of milling. • Proper clamping of workpiece is necessary.
Climb Milling	• Good for slender parts. • Not used for hot-worked metals, forgings, and castings.
Form Milling	• Produces curved profiles with cutters of special shapes. • Can be used for cutting gear teeth.
End Milling	• Versatile to produce various profiles and curved surfaces. • Can produce a variety of surfaces at any depth. • Can operate in a range of 20,000–60,000 rpm in some special cases.
Straddle Milling	• Two or more cutters are mounted on an arbor. • Used to machine two parallel surfaces on the workpiece.
Face Milling	• Surface roughness depends on the corner geometry of the insert and feed per tooth. • During operation, the ratio of cutter diameter to the width of cut should not be less than 3:2.

Drilling

Drilling is a cutting process that uses a drill bit to cut or enlarge a hole of circular cross-section in solid materials.

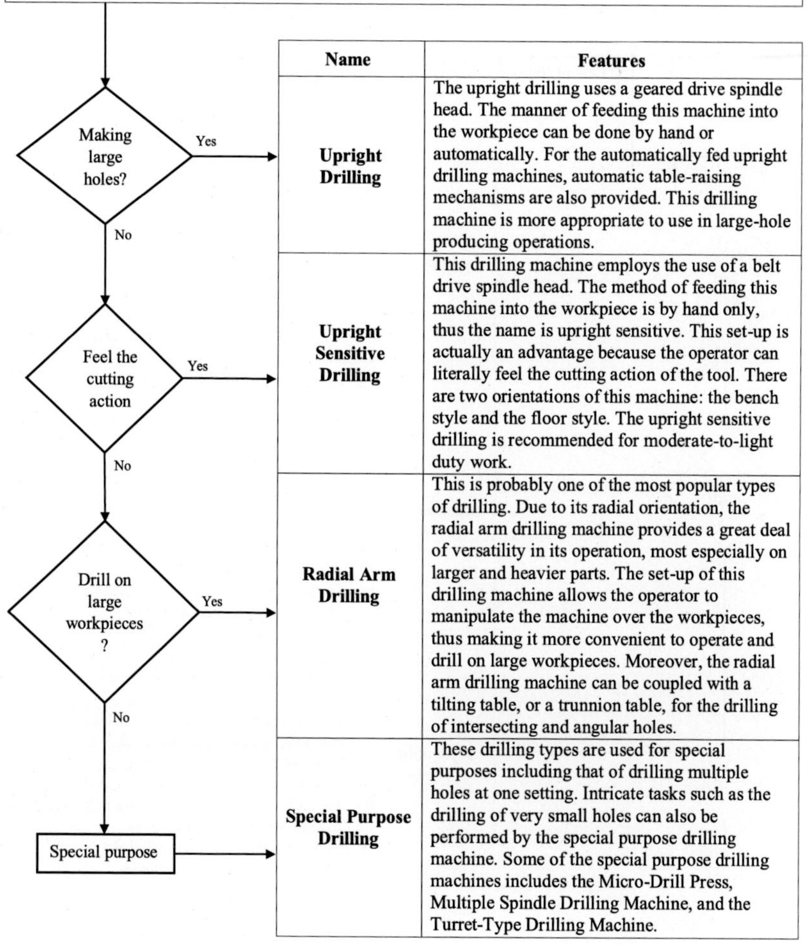

Name	Features
Upright Drilling	The upright drilling uses a geared drive spindle head. The manner of feeding this machine into the workpiece can be done by hand or automatically. For the automatically fed upright drilling machines, automatic table-raising mechanisms are also provided. This drilling machine is more appropriate to use in large-hole producing operations.
Upright Sensitive Drilling	This drilling machine employs the use of a belt drive spindle head. The method of feeding this machine into the workpiece is by hand only, thus the name is upright sensitive. This set-up is actually an advantage because the operator can literally feel the cutting action of the tool. There are two orientations of this machine: the bench style and the floor style. The upright sensitive drilling is recommended for moderate-to-light duty work.
Radial Arm Drilling	This is probably one of the most popular types of drilling. Due to its radial orientation, the radial arm drilling machine provides a great deal of versatility in its operation, most especially on larger and heavier parts. The set-up of this drilling machine allows the operator to manipulate the machine over the workpieces, thus making it more convenient to operate and drill on large workpieces. Moreover, the radial arm drilling machine can be coupled with a tilting table, or a trunnion table, for the drilling of intersecting and angular holes.
Special Purpose Drilling	These drilling types are used for special purposes including that of drilling multiple holes at one setting. Intricate tasks such as the drilling of very small holes can also be performed by the special purpose drilling machine. Some of the special purpose drilling machines includes the Micro-Drill Press, Multiple Spindle Drilling Machine, and the Turret-Type Drilling Machine.

Forming Press

A forming press, commonly shortened to press, is a machine tool that changes the shape of a workpiece by the application of pressure. In addition, some press machine can punch holes in materials. Presses can be classified according to their mechanism, their function and their controllability:

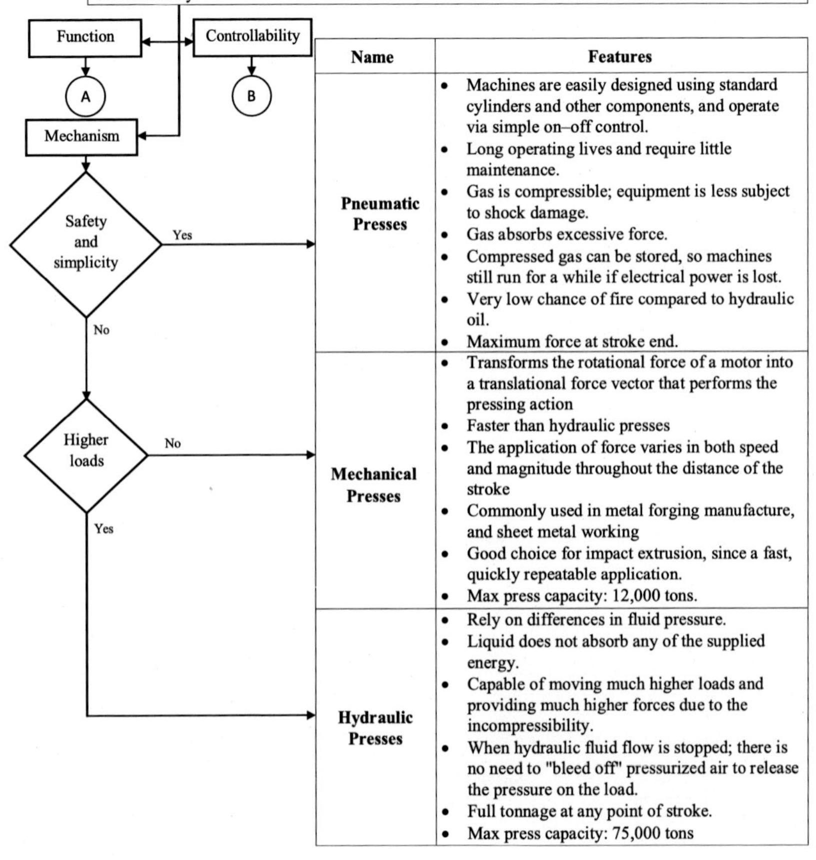

Name	Features
Pneumatic Presses	• Machines are easily designed using standard cylinders and other components, and operate via simple on–off control. • Long operating lives and require little maintenance. • Gas is compressible; equipment is less subject to shock damage. • Gas absorbs excessive force. • Compressed gas can be stored, so machines still run for a while if electrical power is lost. • Very low chance of fire compared to hydraulic oil. • Maximum force at stroke end.
Mechanical Presses	• Transforms the rotational force of a motor into a translational force vector that performs the pressing action • Faster than hydraulic presses • The application of force varies in both speed and magnitude throughout the distance of the stroke • Commonly used in metal forging manufacture, and sheet metal working • Good choice for impact extrusion, since a fast, quickly repeatable application. • Max press capacity: 12,000 tons.
Hydraulic Presses	• Rely on differences in fluid pressure. • Liquid does not absorb any of the supplied energy. • Capable of moving much higher loads and providing much higher forces due to the incompressibility. • When hydraulic fluid flow is stopped; there is no need to "bleed off" pressurized air to release the pressure on the load. • Full tonnage at any point of stroke. • Max press capacity: 75,000 tons

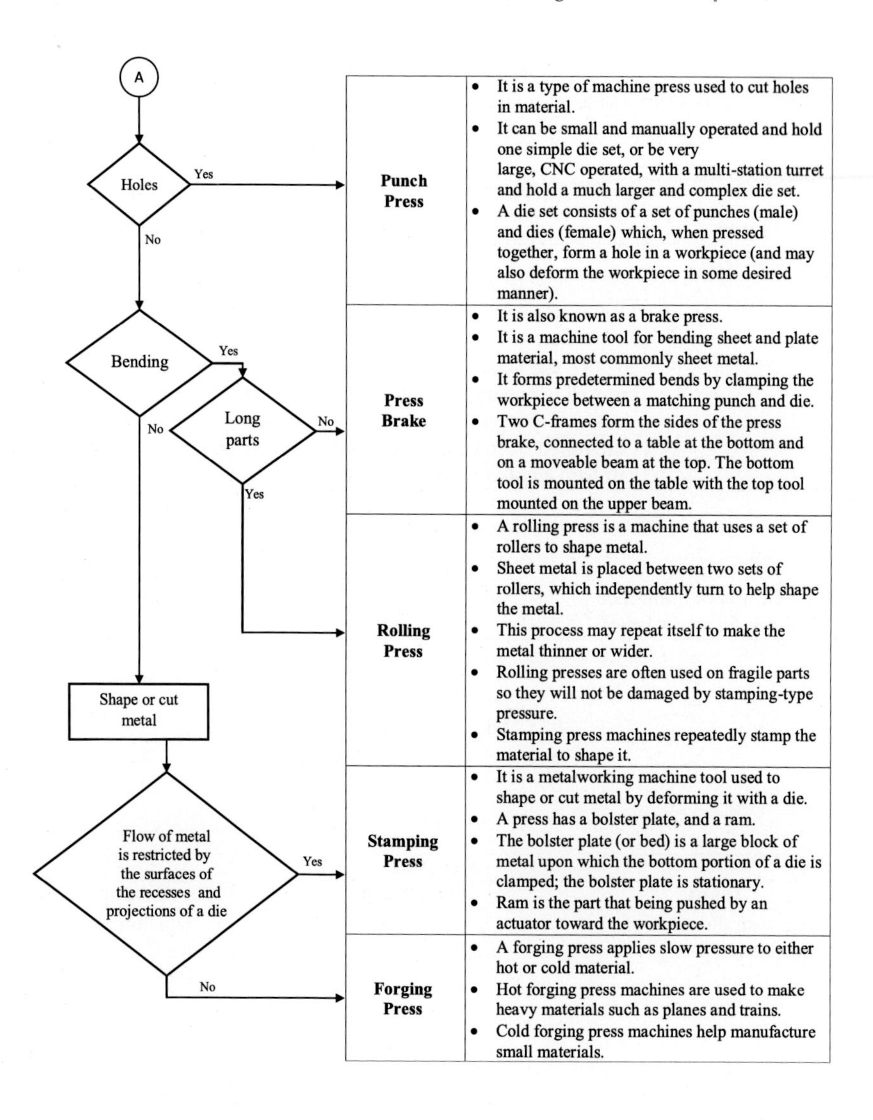
A
Holes
Yes
No
Punch Press
It is a type of machine press used to cut holes in material.
It can be small and manually operated and hold one simple die set, or be very large, CNC operated, with a multi-station turret and hold a much larger and complex die set.
A die set consists of a set of punches (male) and dies (female) which, when pressed together, form a hole in a workpiece (and may also deform the workpiece in some desired manner).
Bending
Yes
No
Long parts
No
Yes
Press Brake
It is also known as a brake press.
It is a machine tool for bending sheet and plate material, most commonly sheet metal.
It forms predetermined bends by clamping the workpiece between a matching punch and die.
Two C-frames form the sides of the press brake, connected to a table at the bottom and on a moveable beam at the top. The bottom tool is mounted on the table with the top tool mounted on the upper beam.
Rolling Press
A rolling press is a machine that uses a set of rollers to shape metal.
Sheet metal is placed between two sets of rollers, which independently turn to help shape the metal.
This process may repeat itself to make the metal thinner or wider.
Rolling presses are often used on fragile parts so they will not be damaged by stamping-type pressure.
Stamping press machines repeatedly stamp the material to shape it.
Shape or cut metal
Flow of metal is restricted by the surfaces of the recesses and projections of a die
Yes
No
Stamping Press
It is a metalworking machine tool used to shape or cut metal by deforming it with a die.
A press has a bolster plate, and a ram.
The bolster plate (or bed) is a large block of metal upon which the bottom portion of a die is clamped; the bolster plate is stationary.
Ram is the part that being pushed by an actuator toward the workpiece.
Forging Press
A forging press applies slow pressure to either hot or cold material.
Hot forging press machines are used to make heavy materials such as planes and trains.
Cold forging press machines help manufacture small materials.

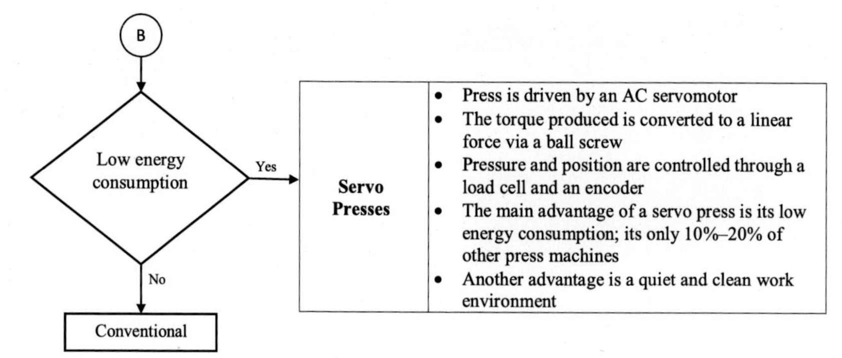
B
Low energy consumption
Yes
No
Conventional
Servo Presses
Press is driven by an AC servomotor
The torque produced is converted to a linear force via a ball screw
Pressure and position are controlled through a load cell and an encoder
The main advantage of a servo press is its low energy consumption; its only 10%–20% of other press machines
Another advantage is a quiet and clean work environment

Shearing Process

Shear stress is a stress state where the stress is parallel to the surface of the material, as opposed to normal stress when the stress is vertical to the surface.

Step 1			
Input	**Formula(s)**		**Output**
Material type	Clearance allowances value		A_c
	Metal group	A_c	
	1100S and 5052S aluminum alloys, all temperatures.	0.045	
	2024ST and 6061ST aluminum alloys; brass all temperatures. Soft cold rolled steel, soft stainless steel.	0.06	
	Cold rolled steel. Half hard: stainless steel. half hard and full hard.	0.075	

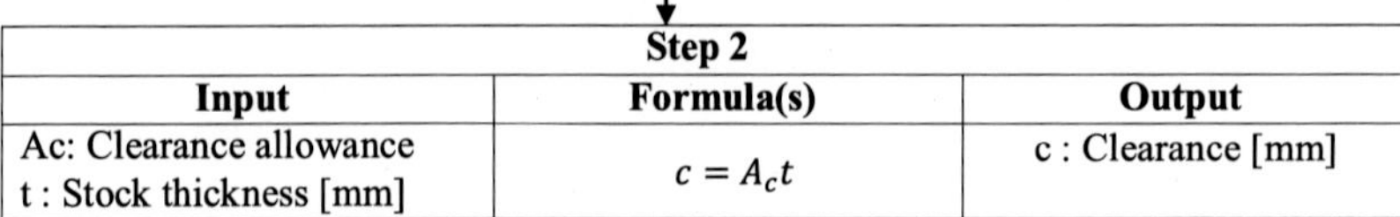

Step 2		
Input	**Formula(s)**	**Output**
Ac: Clearance allowance t : Stock thickness [mm]	$c = A_c t$	c : Clearance [mm]

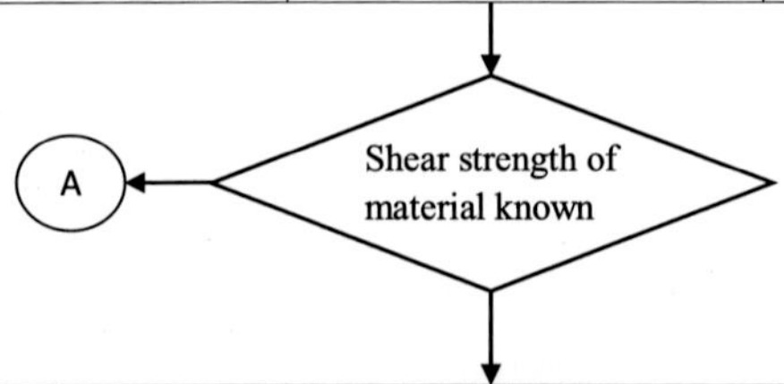

Step 3-a1		
Input	**Formula(s)**	**Output**
Stress–strain graph of the material	Actual Rupture Strength U - Ultimate Strength R – Rupture Strength Y – Yield Point E – Elastic Limit P – Proportional Limit Stress, σ O Strain, ε	U: Ultimate tensile strength [MPa]

B

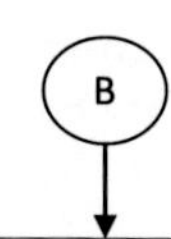

Step 3-a2		
Input	**Formula(s)**	**Output**
U : Ultimate tensile strength of material [MPa] t : Stock thickness [mm] L : Length of the cut edge [mm]	$F = 0.7\,U\,t\,L$	F : Cutting force [N]

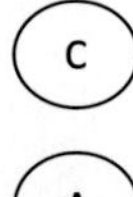

Step 3-b		
Input	**Formula(s)**	**Output**
S: Shear strength of material [MPa] t: Stock thickness [mm] L: Length of the cut edge [mm]	$F = S\,t\,L$	F: Cutting force [N]

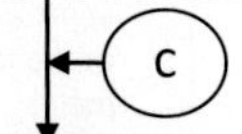

Step 4		
Input	**Formula(s)**	**Output**
D_b: Blanking diameter or D_h: Hole diameter c : Clearance	Blanking punch diameter = D_b -2c. Blanking die diameter = D_b. Hole punch diameter = D_h. Hole die diameter = D_h + 2c.	Blanking punch diameter and Blanking die diameter. or Hole punch diameter and hole die diameter.

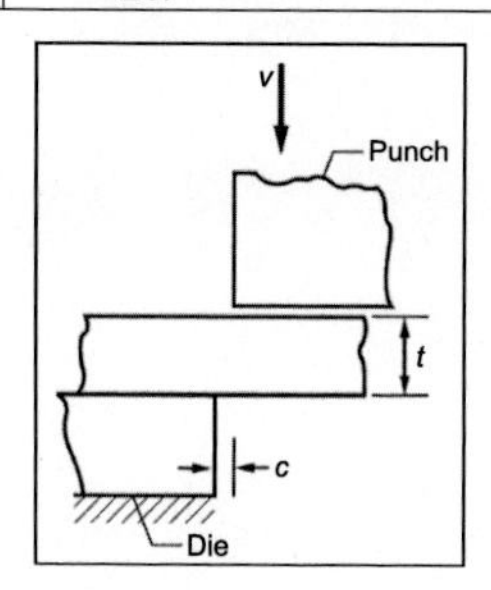

Extrusion

Extrusion is a process used to create objects of a fixed cross-sectional profile. A material is pushed through a die of the desired cross section.

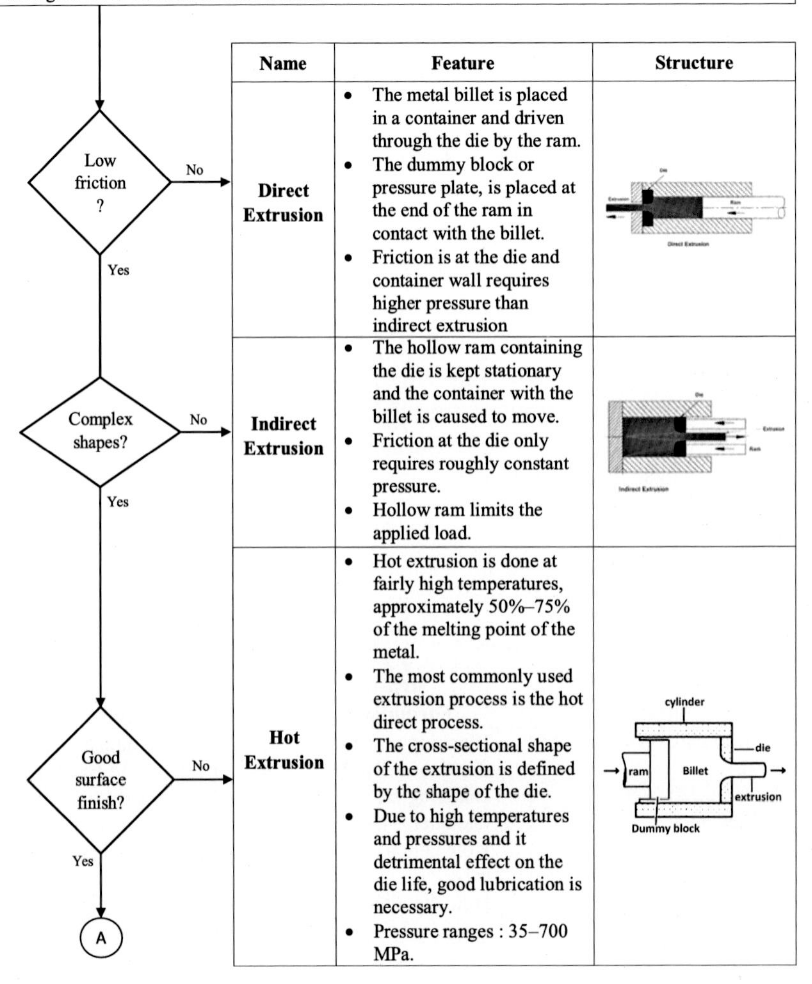

Name	Feature	Structure
Direct Extrusion	• The metal billet is placed in a container and driven through the die by the ram. • The dummy block or pressure plate, is placed at the end of the ram in contact with the billet. • Friction is at the die and container wall requires higher pressure than indirect extrusion	
Indirect Extrusion	• The hollow ram containing the die is kept stationary and the container with the billet is caused to move. • Friction at the die only requires roughly constant pressure. • Hollow ram limits the applied load.	
Hot Extrusion	• Hot extrusion is done at fairly high temperatures, approximately 50%–75% of the melting point of the metal. • The most commonly used extrusion process is the hot direct process. • The cross-sectional shape of the extrusion is defined by the shape of the die. • Due to high temperatures and pressures and it detrimental effect on the die life, good lubrication is necessary. • Pressure ranges : 35–700 MPa.	

A →

Cold Extrusion	• Cold extrusion is the process done at room temperature or slightly elevated temperatures. • This process can be used for materials that can withstand the stresses created by extrusion • No oxidation takes place. • Provide good surface finish with the use of proper lubricants. • Provide good control of dimensional tolerances, no need for heating billet. • Limited deformation can be obtained. • Tooling cost is high.	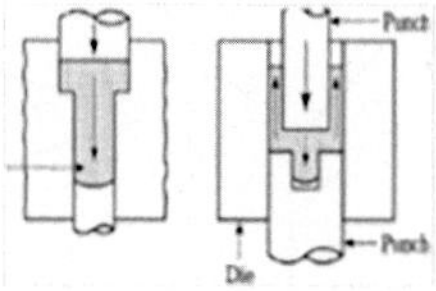

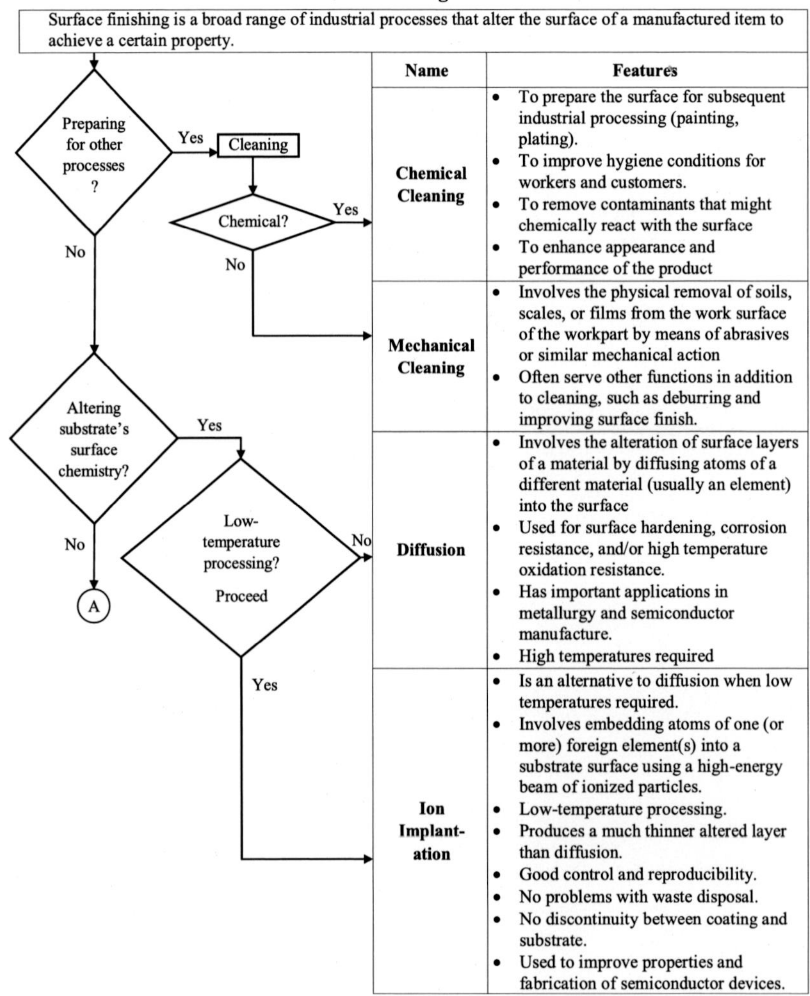

Surface Finishing Methods

Surface finishing is a broad range of industrial processes that alter the surface of a manufactured item to achieve a certain property.

Name	Features
Chemical Cleaning	• To prepare the surface for subsequent industrial processing (painting, plating). • To improve hygiene conditions for workers and customers. • To remove contaminants that might chemically react with the surface • To enhance appearance and performance of the product
Mechanical Cleaning	• Involves the physical removal of soils, scales, or films from the work surface of the workpart by means of abrasives or similar mechanical action • Often serve other functions in addition to cleaning, such as deburring and improving surface finish.
Diffusion	• Involves the alteration of surface layers of a material by diffusing atoms of a different material (usually an element) into the surface • Used for surface hardening, corrosion resistance, and/or high temperature oxidation resistance. • Has important applications in metallurgy and semiconductor manufacture. • High temperatures required
Ion Implant-ation	• Is an alternative to diffusion when low temperatures required. • Involves embedding atoms of one (or more) foreign element(s) into a substrate surface using a high-energy beam of ionized particles. • Low-temperature processing. • Produces a much thinner altered layer than diffusion. • Good control and reproducibility. • No problems with waste disposal. • No discontinuity between coating and substrate. • Used to improve properties and fabrication of semiconductor devices.

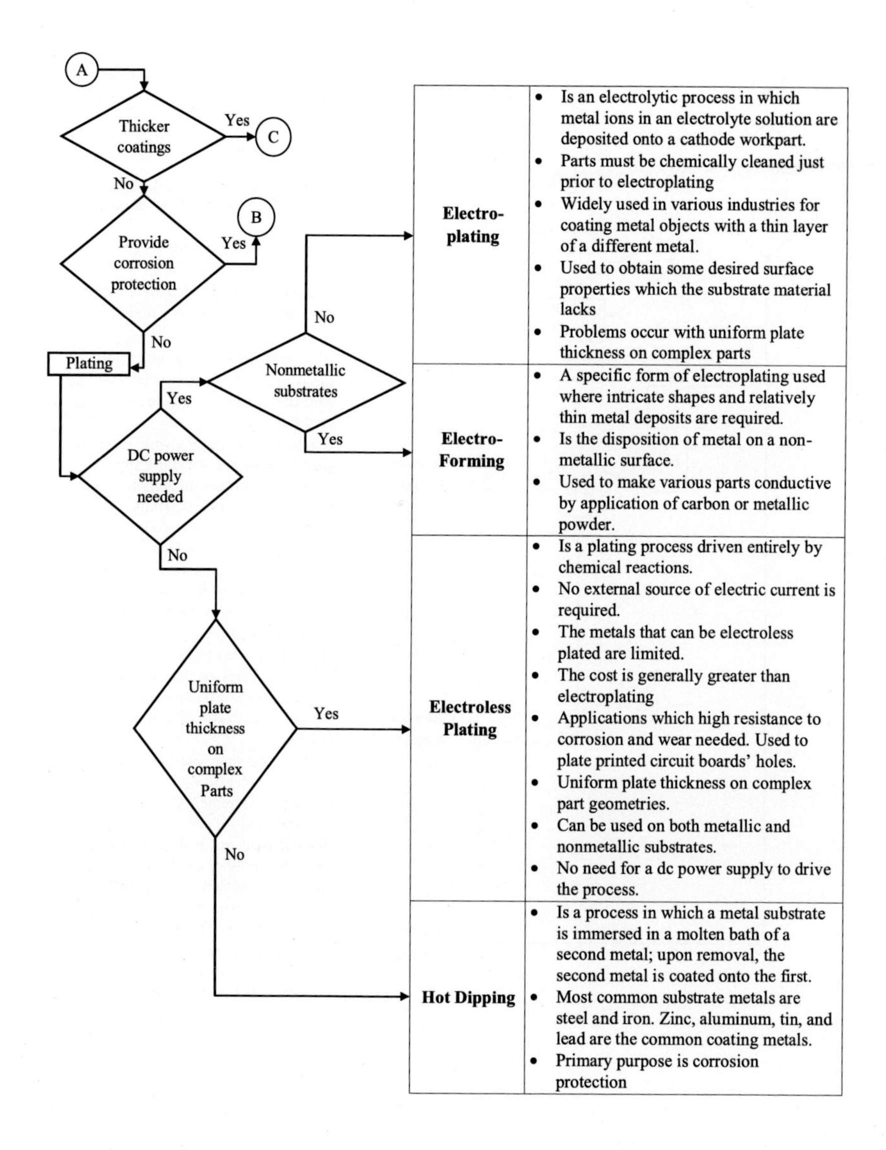
A
Thicker coatings
Yes
C
No
B
Provide corrosion protection
Yes
No
Plating
Nonmetallic substrates
No
Yes
Yes
DC power supply needed
No
Uniform plate thickness on complex Parts
Yes
No
Electro-plating
• Is an electrolytic process in which metal ions in an electrolyte solution are deposited onto a cathode workpart.
• Parts must be chemically cleaned just prior to electroplating
• Widely used in various industries for coating metal objects with a thin layer of a different metal.
• Used to obtain some desired surface properties which the substrate material lacks
• Problems occur with uniform plate thickness on complex parts
Electro-Forming
• A specific form of electroplating used where intricate shapes and relatively thin metal deposits are required.
• Is the disposition of metal on a non-metallic surface.
• Used to make various parts conductive by application of carbon or metallic powder.
Electroless Plating
• Is a plating process driven entirely by chemical reactions.
• No external source of electric current is required.
• The metals that can be electroless plated are limited.
• The cost is generally greater than electroplating
• Applications which high resistance to corrosion and wear needed. Used to plate printed circuit boards' holes.
• Uniform plate thickness on complex part geometries.
• Can be used on both metallic and nonmetallic substrates.
• No need for a dc power supply to drive the process.
Hot Dipping
• Is a process in which a metal substrate is immersed in a molten bath of a second metal; upon removal, the second metal is coated onto the first.
• Most common substrate metals are steel and iron. Zinc, aluminum, tin, and lead are the common coating metals.
• Primary purpose is corrosion protection

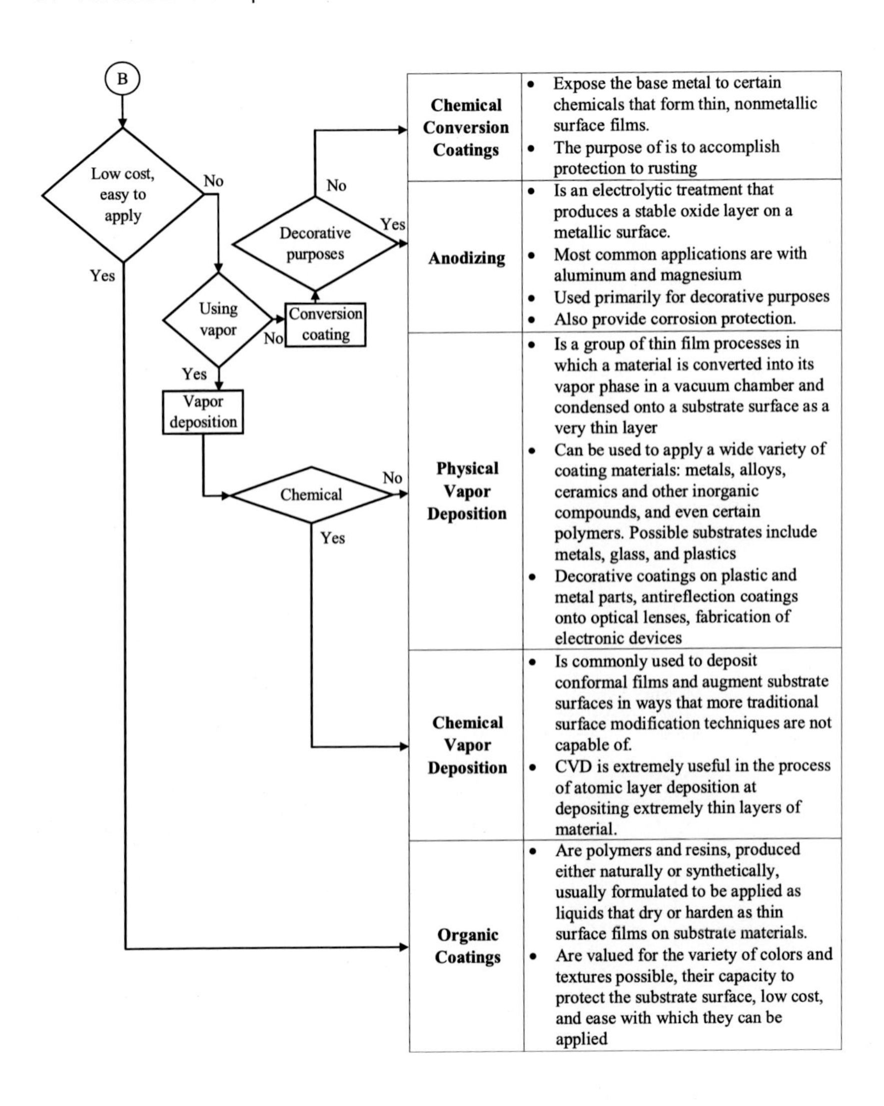
B
Low cost, easy to apply
No
Yes
No
Decorative purposes
Yes
Using vapor
No
Conversion coating
Yes
Vapor deposition
Chemical
No
Yes
Chemical Conversion Coatings
Expose the base metal to certain chemicals that form thin, nonmetallic surface films.
The purpose of is to accomplish protection to rusting
Anodizing
Is an electrolytic treatment that produces a stable oxide layer on a metallic surface.
Most common applications are with aluminum and magnesium
Used primarily for decorative purposes
Also provide corrosion protection.
Physical Vapor Deposition
Is a group of thin film processes in which a material is converted into its vapor phase in a vacuum chamber and condensed onto a substrate surface as a very thin layer
Can be used to apply a wide variety of coating materials: metals, alloys, ceramics and other inorganic compounds, and even certain polymers. Possible substrates include metals, glass, and plastics
Decorative coatings on plastic and metal parts, antireflection coatings onto optical lenses, fabrication of electronic devices
Chemical Vapor Deposition
Is commonly used to deposit conformal films and augment substrate surfaces in ways that more traditional surface modification techniques are not capable of.
CVD is extremely useful in the process of atomic layer deposition at depositing extremely thin layers of material.
Organic Coatings
Are polymers and resins, produced either naturally or synthetically, usually formulated to be applied as liquids that dry or harden as thin surface films on substrate materials.
Are valued for the variety of colors and textures possible, their capacity to protect the substrate surface, low cost, and ease with which they can be applied

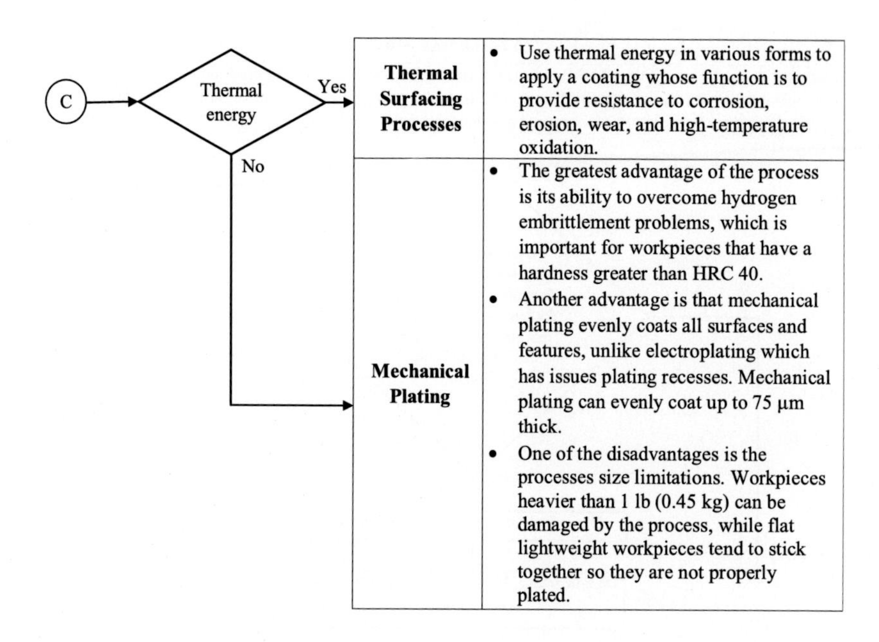
C
Thermal
energy
Yes
No
Thermal
Surfacing
Processes
Use thermal energy in various forms to apply a coating whose function is to provide resistance to corrosion, erosion, wear, and high-temperature oxidation.
Mechanical
Plating
The greatest advantage of the process is its ability to overcome hydrogen embrittlement problems, which is important for workpieces that have a hardness greater than HRC 40.
Another advantage is that mechanical plating evenly coats all surfaces and features, unlike electroplating which has issues plating recesses. Mechanical plating can evenly coat up to 75 μm thick.
One of the disadvantages is the processes size limitations. Workpieces heavier than 1 lb (0.45 kg) can be damaged by the process, while flat lightweight workpieces tend to stick together so they are not properly plated.

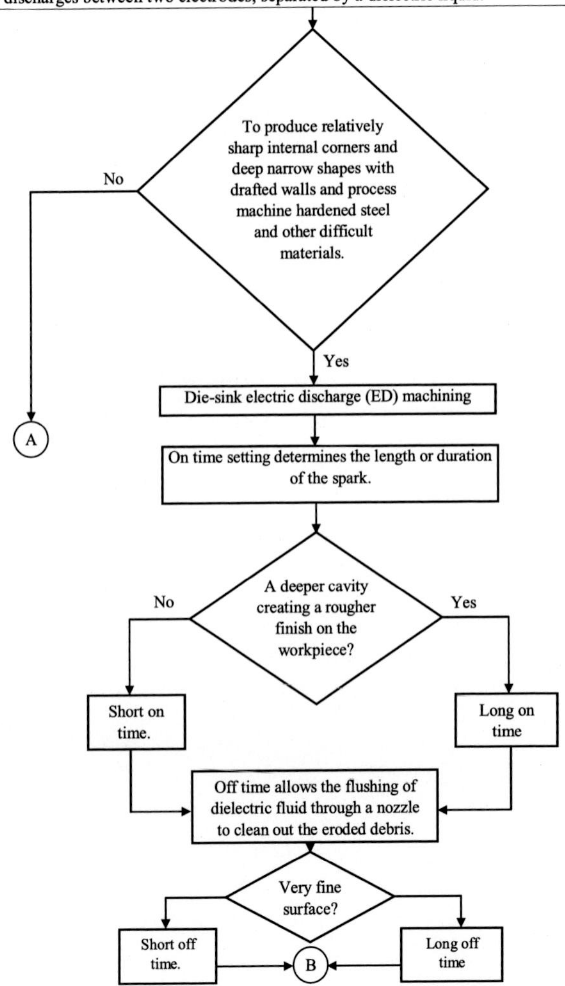
Electro Discharge Machining
Electrical discharge machining (EDM) is a manufacturing process whereby a desired shape is obtained using electrical discharges (sparks). Material is removed from the workpiece by a series of rapidly recurring current discharges between two electrodes, separated by a dielectric liquid.
To produce relatively sharp internal corners and deep narrow shapes with drafted walls and process machine hardened steel and other difficult materials.
No
Yes
A
Die-sink electric discharge (ED) machining
On time setting determines the length or duration of the spark.
A deeper cavity creating a rougher finish on the workpiece?
No
Yes
Short on time.
Long on time
Off time allows the flushing of dielectric fluid through a nozzle to clean out the eroded debris.
Very fine surface?
Short off time.
B
Long off time

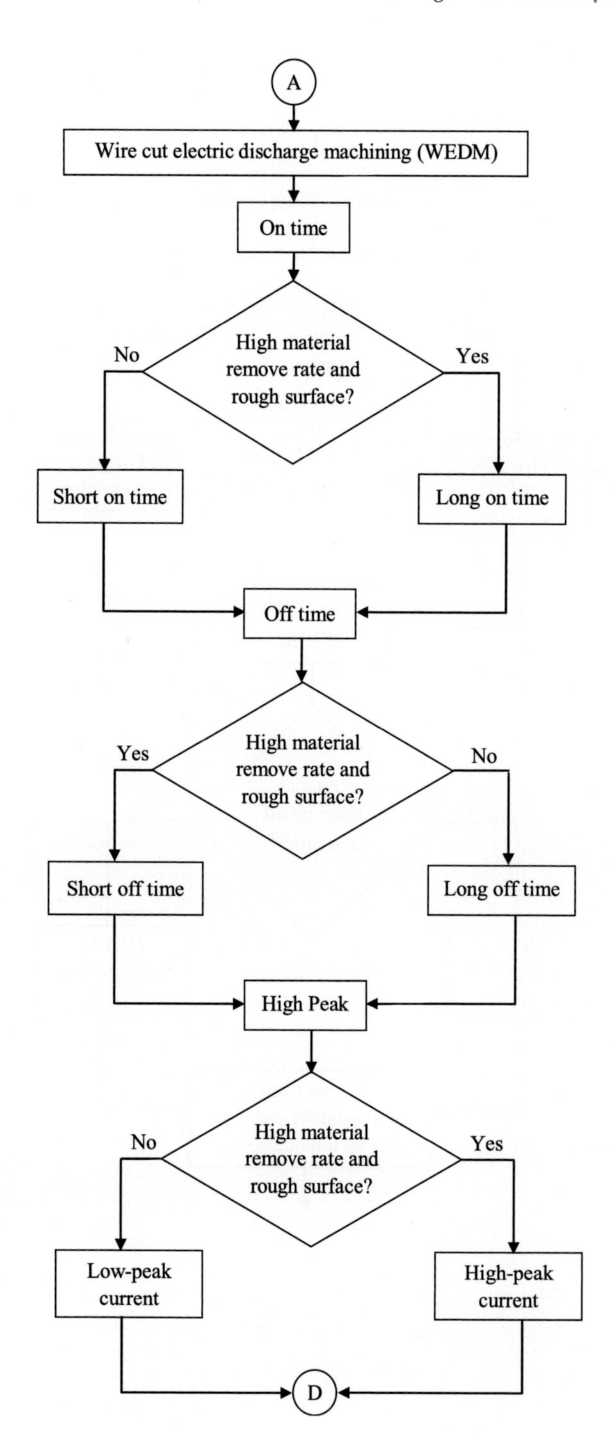
A
Wire cut electric discharge machining (WEDM)
On time
High material remove rate and rough surface?
No
Yes
Short on time
Long on time
Off time
High material remove rate and rough surface?
Yes
No
Short off time
Long off time
High Peak
High material remove rate and rough surface?
No
Yes
Low-peak current
High-peak current
D

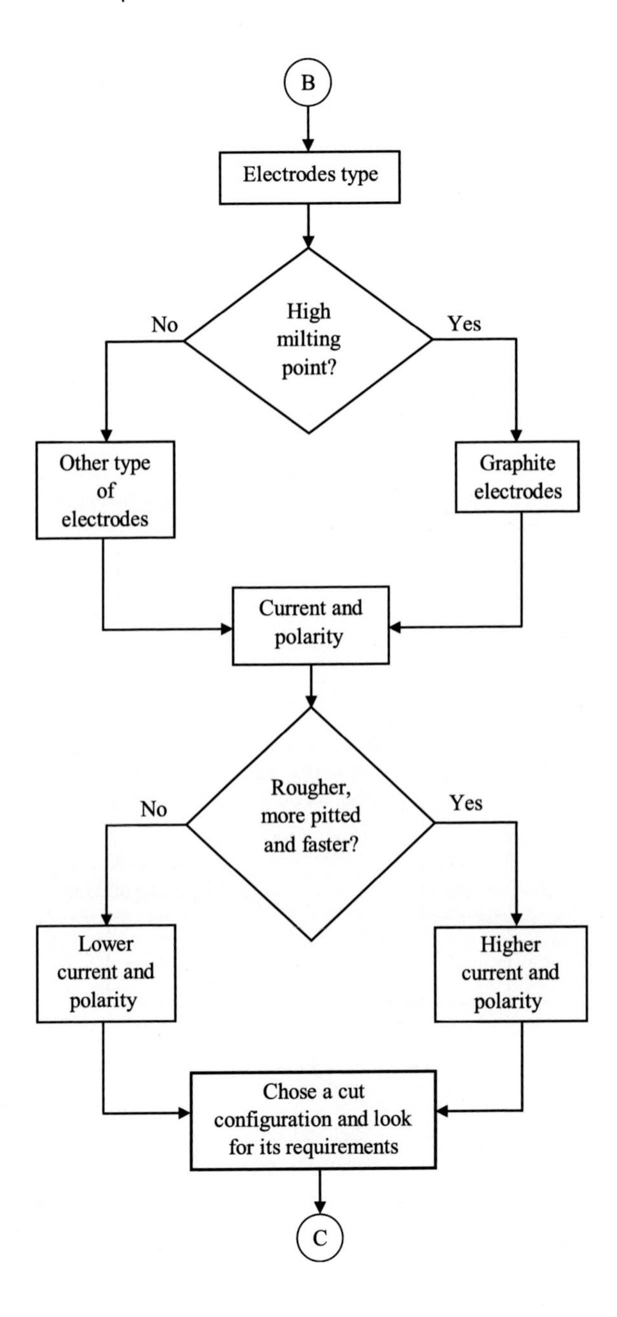
B
Electrodes type
High
milting
point?
No
Yes
Other type
of
electrodes
Graphite
electrodes
Current and
polarity
Rougher,
more pitted
and faster?
No
Yes
Lower
current and
polarity
Higher
current and
polarity
Chose a cut
configuration and look
for its requirements
C

Name	Featrues	Structure
Deep small radius holes and grooves.	The electrode moves with a slow downward movement, workpiece is fixed.	Electrode Dielectric fluid Downward movement Workpiece
More complicated configuration.	Either the electrode or the workpiece moves two-dimensionally. two-axes machining.	Electrode Workpiece or electrode 2-dimensional movement Conducting medium Workpiece
If the bottom of the slots in image 2 is contoured in up-and-down directions.	Additional programmable axis in the nachine for three-dimensional movement.	Dielectric fluid Electrode 3-dimensional movement Workpiece
EDM to downward and rotational direction.	One simple electrode, possibly a standard-type electrode and use a rotational axis to make holes and a slot.	Rotational and downward movement Electrode Dielectric fluid Workpiece

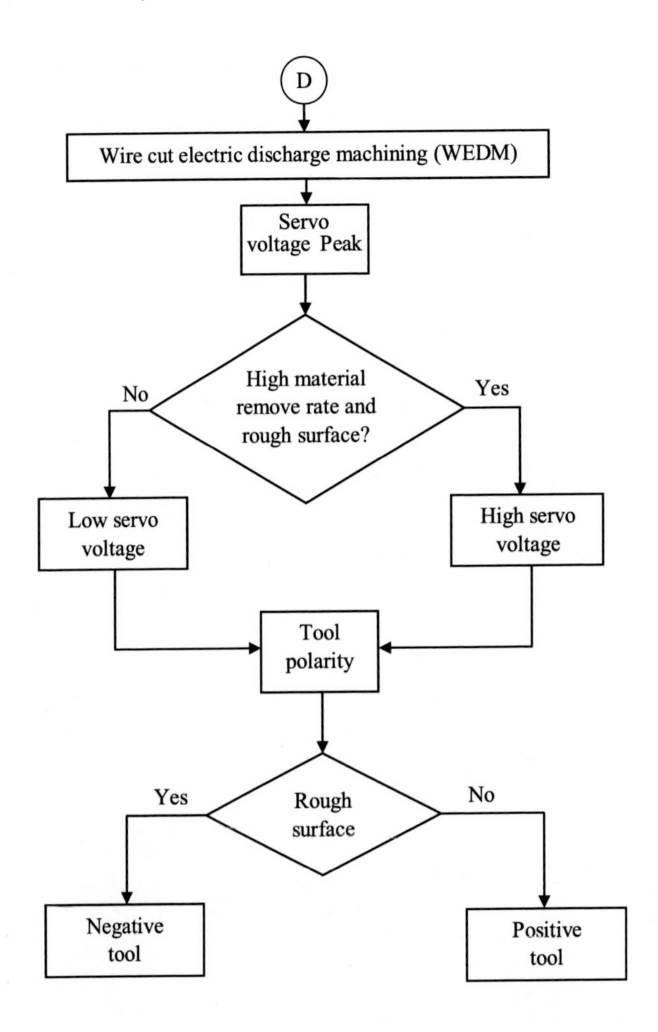

The WEDM configuration can contradict with each other. There is a lot of research about the optimal values for the upper parameters in different situation. Even algorithms were put for this purpose.

Casting

Casting is a process in which molten metarial flows by gravity or other force into a mold where it solidifies in the shape of the mold cavity

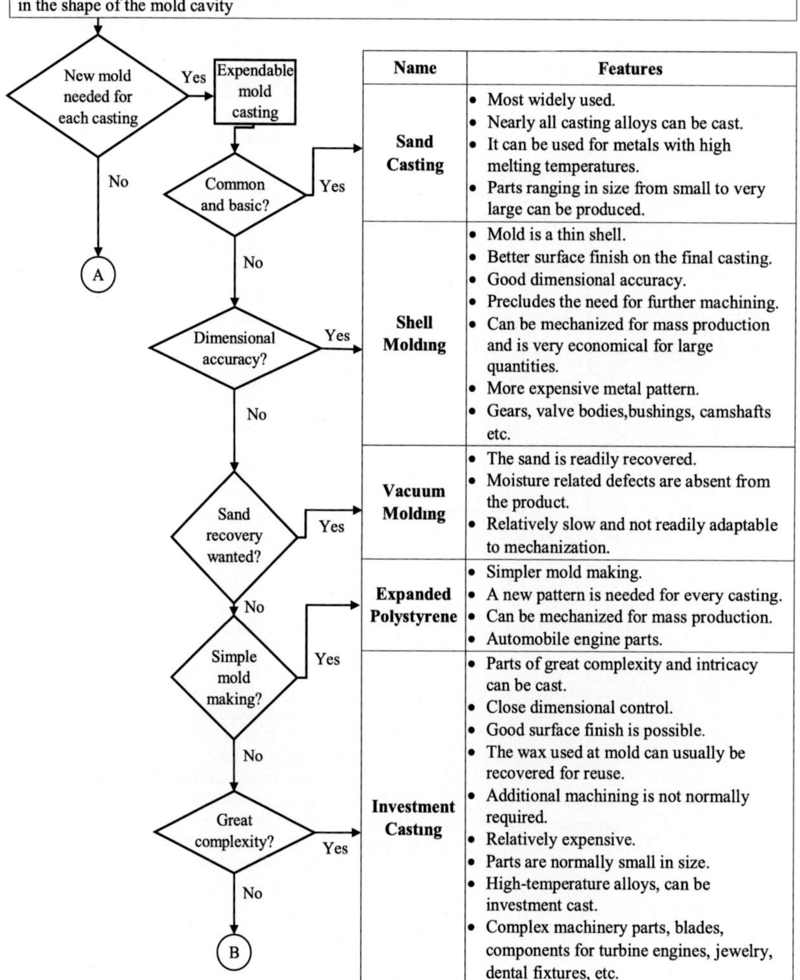

Name	Features
Sand Casting	• Most widely used. • Nearly all casting alloys can be cast. • It can be used for metals with high melting temperatures. • Parts ranging in size from small to very large can be produced.
Shell Moldıng	• Mold is a thin shell. • Better surface finish on the final casting. • Good dimensional accuracy. • Precludes the need for further machining. • Can be mechanized for mass production and is very economical for large quantities. • More expensive metal pattern. • Gears, valve bodies,bushings, camshafts etc.
Vacuum Moldıng	• The sand is readily recovered. • Moisture related defects are absent from the product. • Relatively slow and not readily adaptable to mechanization.
Expanded Polystyrene	• Simpler mold making. • A new pattern is needed for every casting. • Can be mechanized for mass production. • Automobile engine parts.
Investment Castıng	• Parts of great complexity and intricacy can be cast. • Close dimensional control. • Good surface finish is possible. • The wax used at mold can usually be recovered for reuse. • Additional machining is not normally required. • Relatively expensive. • Parts are normally small in size. • High-temperature alloys, can be investment cast. • Complex machinery parts, blades, components for turbine engines, jewelry, dental fixtures, etc.

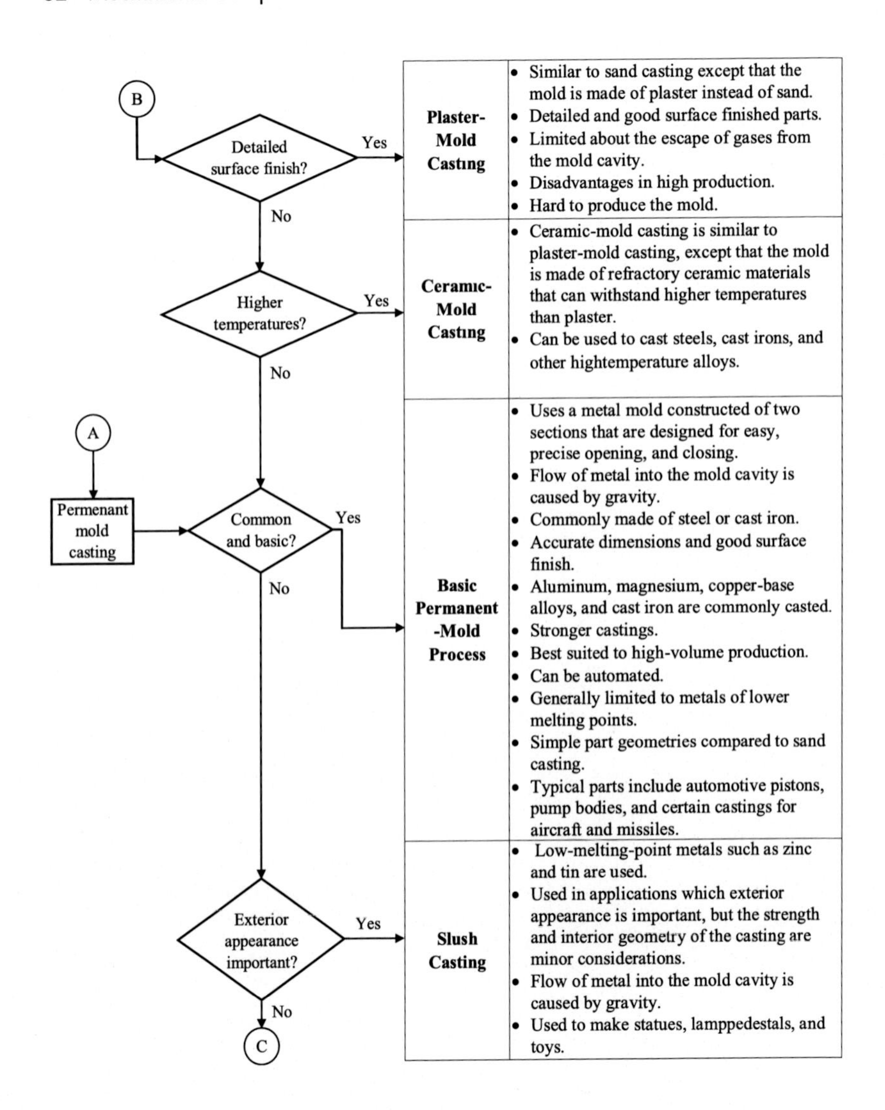
B
Detailed
surface finish?
Yes
No
Higher
temperatures?
Yes
No
A
Permenant
mold
casting
Common
and basic?
Yes
No
Exterior
appearance
important?
Yes
No
C
Plaster-
Mold
Castıng
Similar to sand casting except that the mold is made of plaster instead of sand.
Detailed and good surface finished parts.
Limited about the escape of gases from the mold cavity.
Disadvantages in high production.
Hard to produce the mold.
Ceramıc-
Mold
Castıng
Ceramic-mold casting is similar to plaster-mold casting, except that the mold is made of refractory ceramic materials that can withstand higher temperatures than plaster.
Can be used to cast steels, cast irons, and other hightemperature alloys.
Basic
Permanent
-Mold
Process
Uses a metal mold constructed of two sections that are designed for easy, precise opening, and closing.
Flow of metal into the mold cavity is caused by gravity.
Commonly made of steel or cast iron.
Accurate dimensions and good surface finish.
Aluminum, magnesium, copper-base alloys, and cast iron are commonly casted.
Stronger castings.
Best suited to high-volume production.
Can be automated.
Generally limited to metals of lower melting points.
Simple part geometries compared to sand casting.
Typical parts include automotive pistons, pump bodies, and certain castings for aircraft and missiles.
Slush
Casting
Low-melting-point metals such as zinc and tin are used.
Used in applications which exterior appearance is important, but the strength and interior geometry of the casting are minor considerations.
Flow of metal into the mold cavity is caused by gravity.
Used to make statues, lamppedestals, and toys.

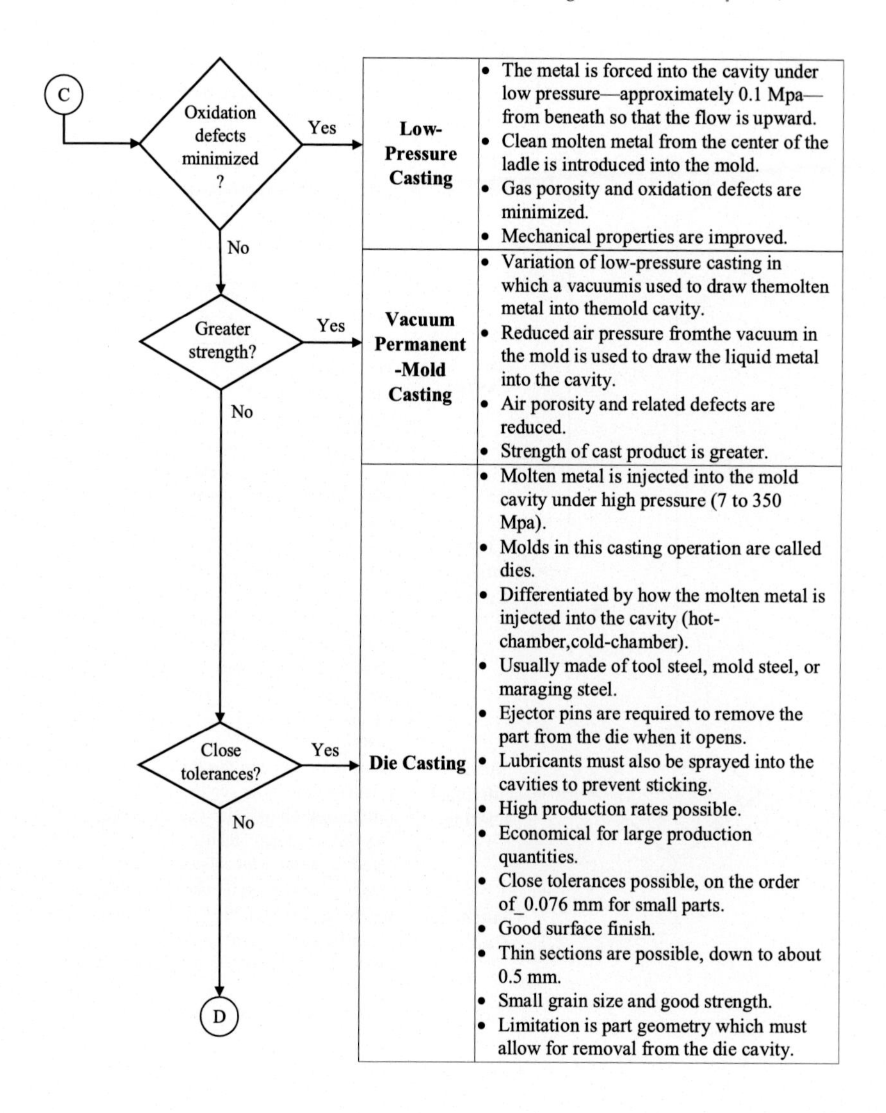
C
Oxidation defects minimized ?
Yes
No
Low-Pressure Casting
• The metal is forced into the cavity under low pressure—approximately 0.1 Mpa—from beneath so that the flow is upward.
• Clean molten metal from the center of the ladle is introduced into the mold.
• Gas porosity and oxidation defects are minimized.
• Mechanical properties are improved.
Greater strength?
Yes
No
Vacuum Permanent -Mold Casting
• Variation of low-pressure casting in which a vacuumis used to draw themolten metal into themold cavity.
• Reduced air pressure fromthe vacuum in the mold is used to draw the liquid metal into the cavity.
• Air porosity and related defects are reduced.
• Strength of cast product is greater.
Close tolerances?
Yes
No
Die Casting
• Molten metal is injected into the mold cavity under high pressure (7 to 350 Mpa).
• Molds in this casting operation are called dies.
• Differentiated by how the molten metal is injected into the cavity (hot-chamber,cold-chamber).
• Usually made of tool steel, mold steel, or maraging steel.
• Ejector pins are required to remove the part from the die when it opens.
• Lubricants must also be sprayed into the cavities to prevent sticking.
• High production rates possible.
• Economical for large production quantities.
• Close tolerances possible, on the order of_0.076 mm for small parts.
• Good surface finish.
• Thin sections are possible, down to about 0.5 mm.
• Small grain size and good strength.
• Limitation is part geometry which must allow for removal from the die cavity.
D

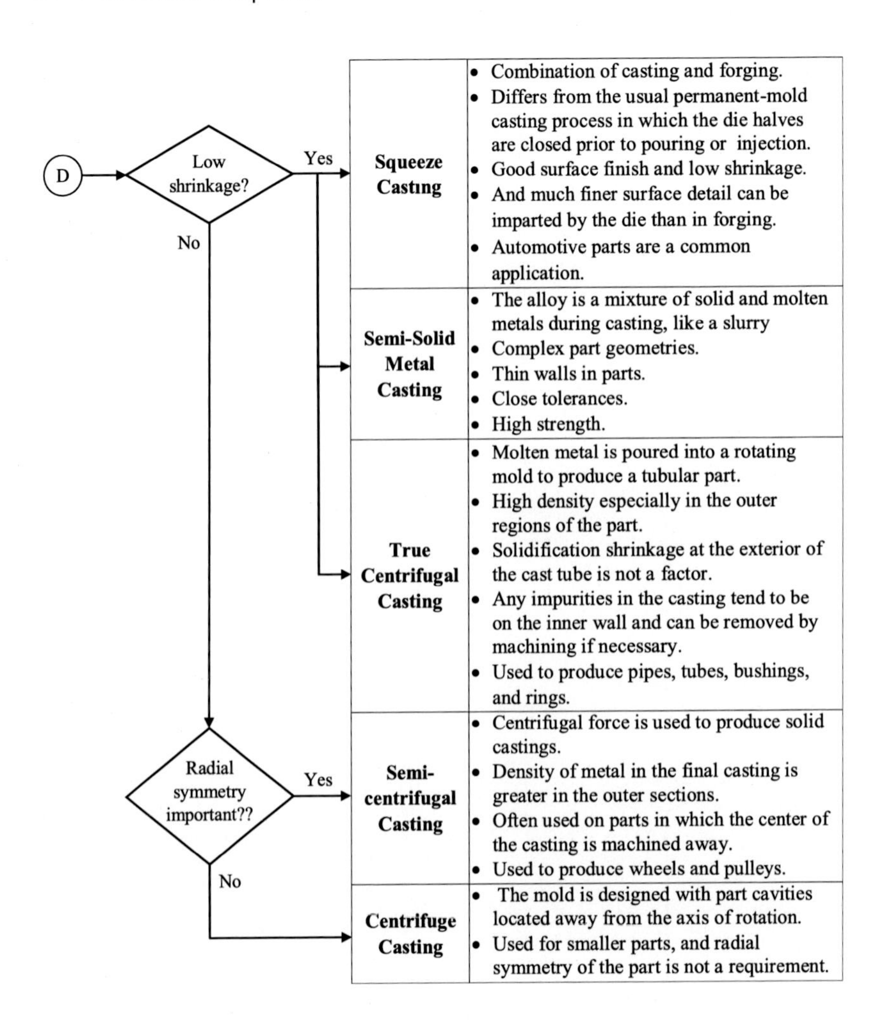
D
Low shrinkage?
Yes
No
Squeeze Casting
• Combination of casting and forging.
• Differs from the usual permanent-mold casting process in which the die halves are closed prior to pouring or injection.
• Good surface finish and low shrinkage.
• And much finer surface detail can be imparted by the die than in forging.
• Automotive parts are a common application.
Semi-Solid Metal Casting
• The alloy is a mixture of solid and molten metals during casting, like a slurry
• Complex part geometries.
• Thin walls in parts.
• Close tolerances.
• High strength.
True Centrifugal Casting
• Molten metal is poured into a rotating mold to produce a tubular part.
• High density especially in the outer regions of the part.
• Solidification shrinkage at the exterior of the cast tube is not a factor.
• Any impurities in the casting tend to be on the inner wall and can be removed by machining if necessary.
• Used to produce pipes, tubes, bushings, and rings.
Radial symmetry important??
Yes
No
Semi-centrifugal Casting
• Centrifugal force is used to produce solid castings.
• Density of metal in the final casting is greater in the outer sections.
• Often used on parts in which the center of the casting is machined away.
• Used to produce wheels and pulleys.
Centrifuge Casting
• The mold is designed with part cavities located away from the axis of rotation.
• Used for smaller parts, and radial symmetry of the part is not a requirement.

Sintering

Sintering is the process of compacting and forming a solid mass of material by heat and/or pressure without melting it to the point of liquefaction.

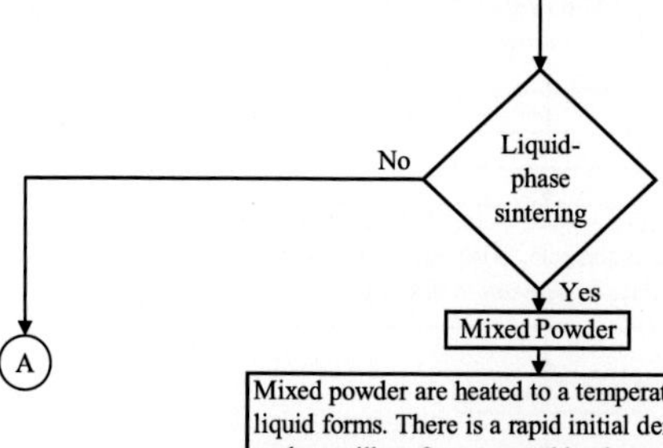

Mixed Powder

Mixed powder are heated to a temperature where a liquid forms. There is a rapid initial densification due to the capillary force exerted by the wetting liquid on the solid particles.

Rearrangement

During rearrengement, the compact response as a viscous solid the capillary action. The elimination of porosity increases the compact viscosity and as a result of this the densification rate continuously decreases. The amount of densification is dependent on the amount of liquid, particle size and solubility of the solid in the liquid.

Solution - Reprecipitation

A general attribute of solution-reprecipitation proceses is microstructural coarsening that is due to a distrubution in grain sizes. Materials is transported from the small grains to the large grain by diffusion. This step not only contrubutes to grain coarsening, but also to densification.

Solid state

The last step of liquid phase sintering is solid state. This step includes pore elimination, grain growth, and contact growt.

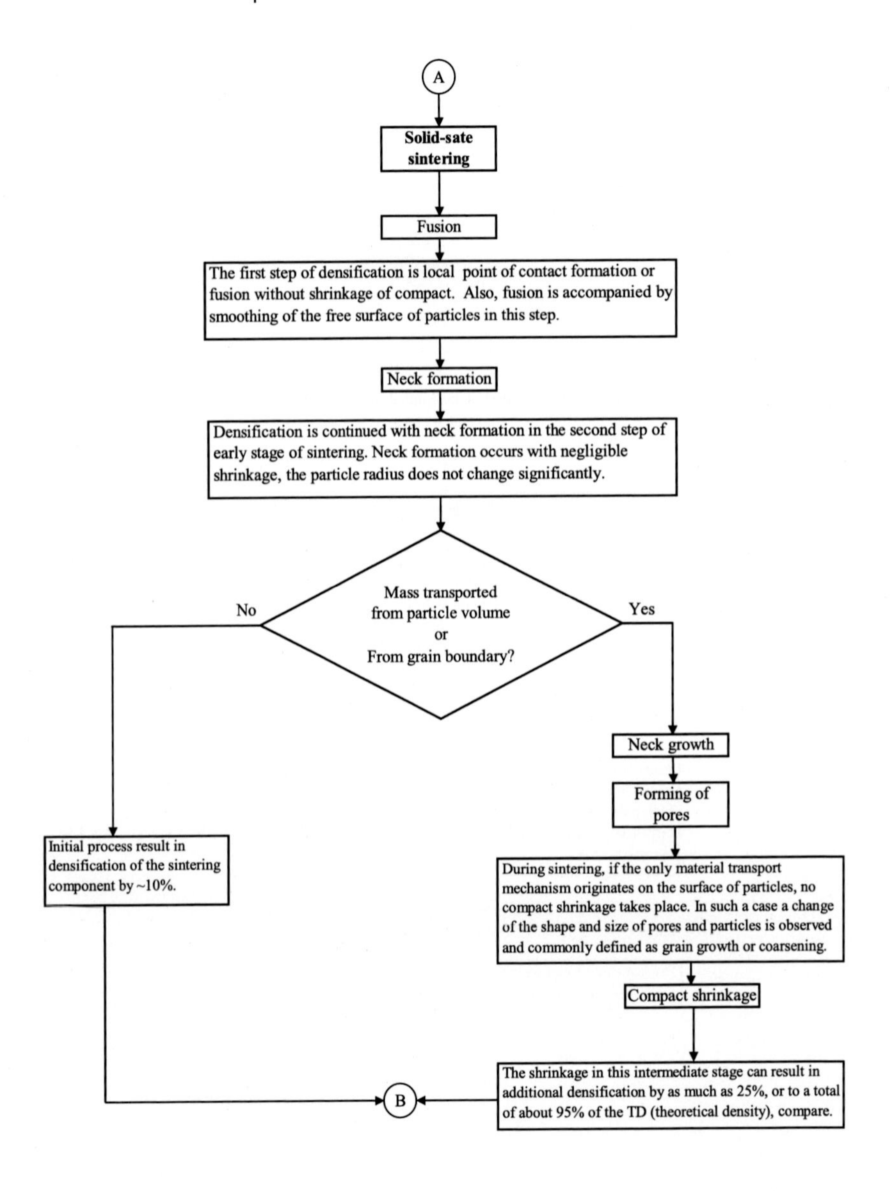
A
Solid-sate sintering
Fusion
The first step of densification is local point of contact formation or fusion without shrinkage of compact. Also, fusion is accompanied by smoothing of the free surface of particles in this step.
Neck formation
Densification is continued with neck formation in the second step of early stage of sintering. Neck formation occurs with negligible shrinkage, the particle radius does not change significantly.
Mass transported from particle volume or From grain boundary?
No
Yes
Initial process result in densification of the sintering component by ~10%.
Neck growth
Forming of pores
During sintering, if the only material transport mechanism originates on the surface of particles, no compact shrinkage takes place. In such a case a change of the shape and size of pores and particles is observed and commonly defined as grain growth or coarsening.
Compact shrinkage
The shrinkage in this intermediate stage can result in additional densification by as much as 25%, or to a total of about 95% of the TD (theoretical density), compare.
B

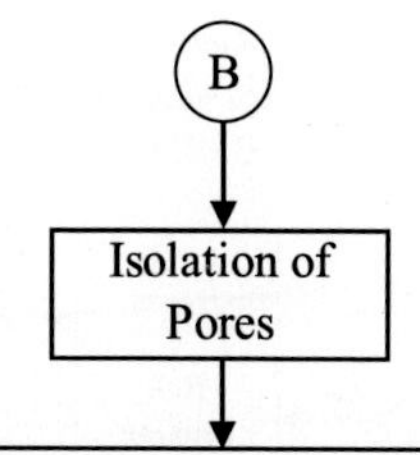

With isolation of pores, the final sintering stage begins at about 93–95% of theoretical density.

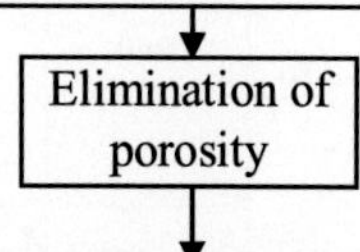

The complete elimination of porosity can only happen when all pores are connected to fast, short diffusion paths along grain boundaries.

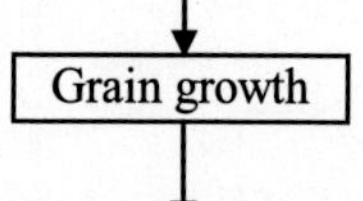

Small pores attached to grain boundaries move quickly to collapse together and thus reduce surface energy. Along with pores, grain boundary movement is accelerated, leading eventually to discontinuous grain growth, when pores do not pin any more the grain boundaries. This leads to rapidly moving grain boundaries that consume small grain

Rolling

Rolling is a metal forming process in which metal stock is passed through one or more pairs of rolls to reduce the thickness and to make the thickness uniform.

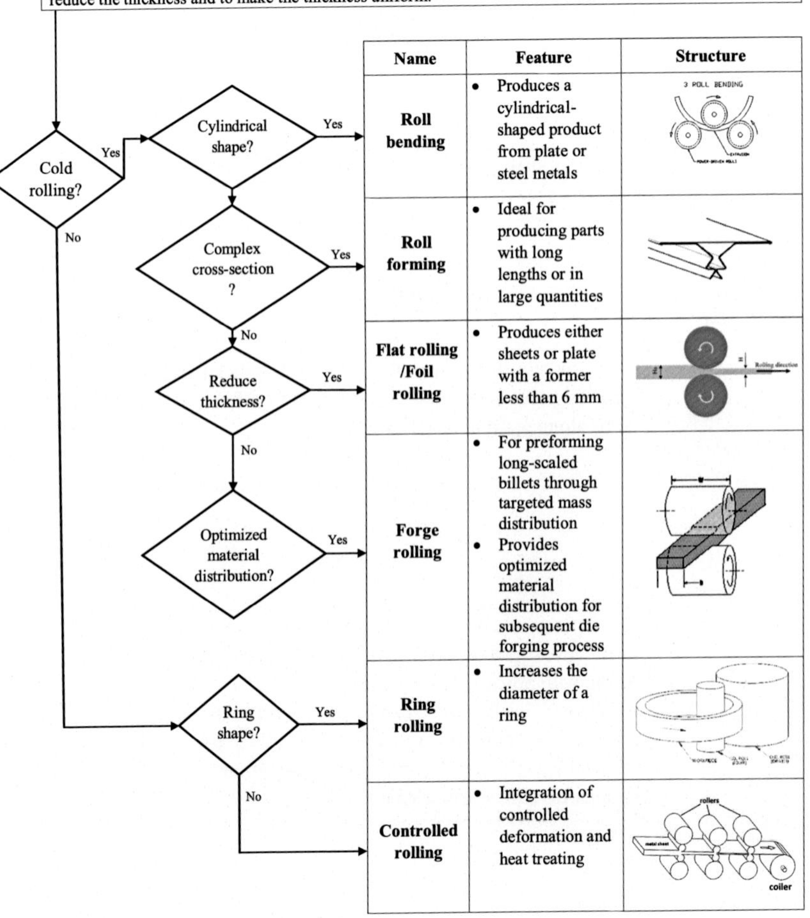

Name	Feature	Structure
Roll bending	• Produces a cylindrical-shaped product from plate or steel metals	
Roll forming	• Ideal for producing parts with long lengths or in large quantities	
Flat rolling /Foil rolling	• Produces either sheets or plate with a former less than 6 mm	
Forge rolling	• For preforming long-scaled billets through targeted mass distribution • Provides optimized material distribution for subsequent die forging process	
Ring rolling	• Increases the diameter of a ring	
Controlled rolling	• Integration of controlled deformation and heat treating	

Forging

Forging is the working of metal into a useful shape by hammering or pressing. Assuming that; upsetting a cylinder with open die forging, metal is cold formed, platen workpiece interface is frictionless.

Step 1

Terms	Formula(s)	Notes
A_0: Initial cross-sectional area (m^2). A_1: Deformed area (m^2). h_0: Original cylinder height (m). h_1: Deformed height (m).	Due to volume equality: $A_0 h_0 = A_1 h_1$	Deformation proceeds the cylinder becomes shorter.

Step 2

Terms	Formula(s)	Notes
K: Material strength coefficient (Pa). ε: True strain (true stress required for upsetting). n: Strain hardening exponent. V: Platen velocity (m/s). h:instantaneous height (m). m: Strain strength exponent. C: Strain rate strength constant.	**Strain rate** For cold forging: $\varepsilon_1 = ln\frac{A_0}{A_1}$ For hot forging: $\varepsilon = \frac{V}{h}$ **Hollomon equation** For cold forging $\sigma_t = K\varepsilon_1^n = K\left[ln\left(\frac{A_0}{A_1}\right)\right]^n$ For hot forging: $\sigma_t = C\dot{\varepsilon}^m$	True stress required for upsetting (often called the flow stress in forming situations).

Step 3

Terms	Formula(s)	Notes	
K: Material strength coefficient (Pa). A: The projected area of the forging including the flash (m^2). J: Multiplying factor. Y_f: Flow stress of material (Pa).	**Forging force** For cold forging: $F = K\left[ln\left(\frac{A_0}{A_1}\right)\right]^n A$ For hot forging: $F = JY_f A$	Range of J	
		Shape	J
		Simple shape without a flash	3–5
		Simple shape with a flash	5–8
		Complex shape with a flash	8–12

(A)

A

Step 4

Terms	Formula(s)	Notes
W: Work (j). V: Volume of the cylinder (m^3). ε_1: Total true strain due to upsetting.	For cold forging: $W = \frac{KV{\varepsilon_1}^{n+1}}{n+1}$ For hot forging: $W = CV\dot{\varepsilon}^m\varepsilon_1$	Ideal work for deformation.

Step 5

Terms	Formula(s)	Notes
P_{av}: Average power per cycle (w). t_{av}: Average time per cycle (s).	For cold forging: $p_{av} = \frac{1}{t_{av}}\left(\frac{KV{\varepsilon_1}^{n+1}}{n+1}\right)$ For hot forging: $p_{av} = \frac{1}{t_{av}}(CV\dot{\varepsilon}^m\varepsilon_1)$	Power consuming per cycle.

Forging Process

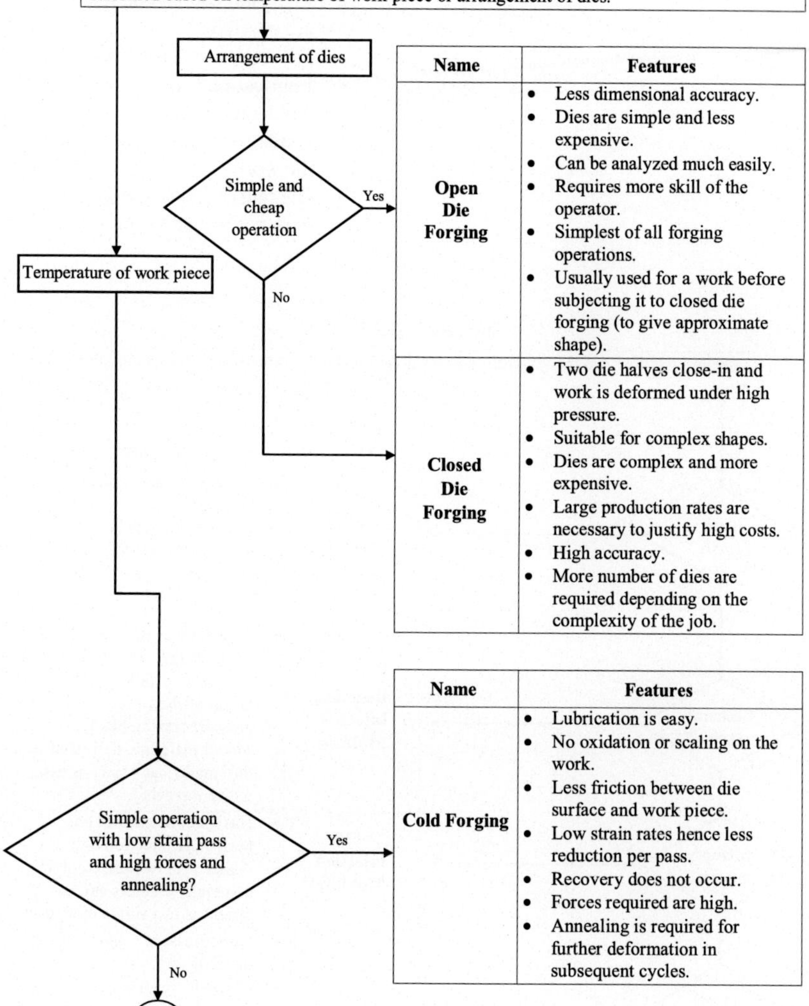

A →

Hot Forging	• High strain rates and hence easy flow of metal. • Recrystallization and recovery are possible. • Forces required are less. • Lubrication is difficult at high temperatures. • Oxidation and scaling occur on the work. • Die must withstand high working temperature.

Injection Molding

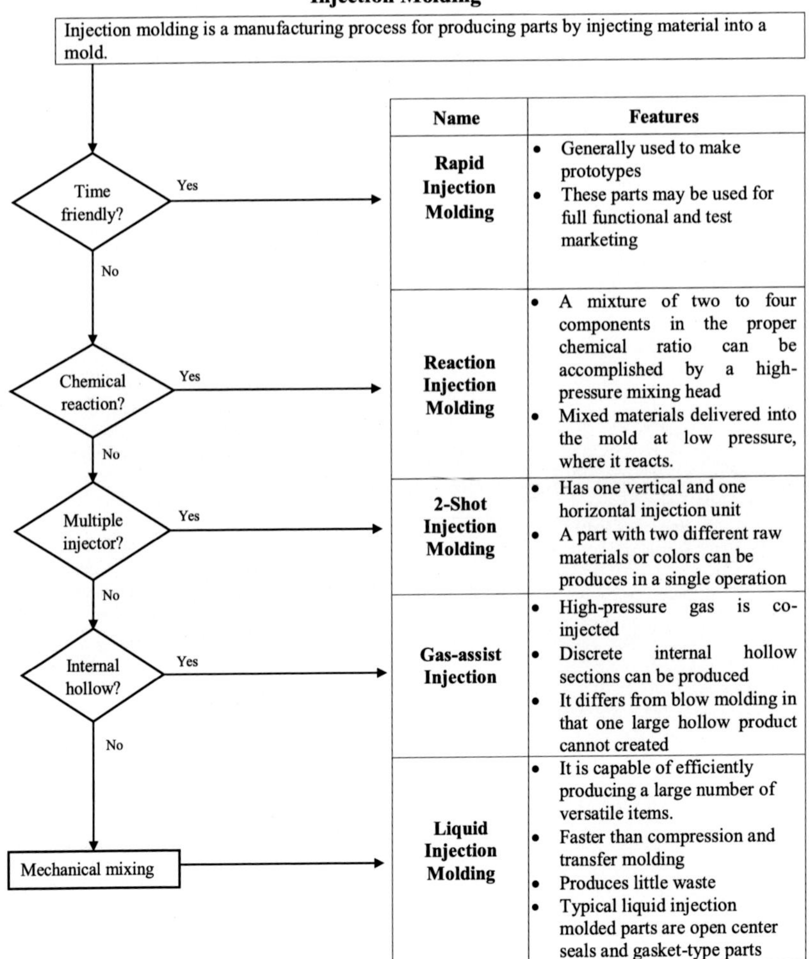

Name	Features
Rapid Injection Molding	• Generally used to make prototypes • These parts may be used for full functional and test marketing
Reaction Injection Molding	• A mixture of two to four components in the proper chemical ratio can be accomplished by a high-pressure mixing head • Mixed materials delivered into the mold at low pressure, where it reacts.
2-Shot Injection Molding	• Has one vertical and one horizontal injection unit • A part with two different raw materials or colors can be produces in a single operation
Gas-assist Injection	• High-pressure gas is co-injected • Discrete internal hollow sections can be produced • It differs from blow molding in that one large hollow product cannot created
Liquid Injection Molding	• It is capable of efficiently producing a large number of versatile items. • Faster than compression and transfer molding • Produces little waste • Typical liquid injection molded parts are open center seals and gasket-type parts

Hot End / Nozzle Design Procedure

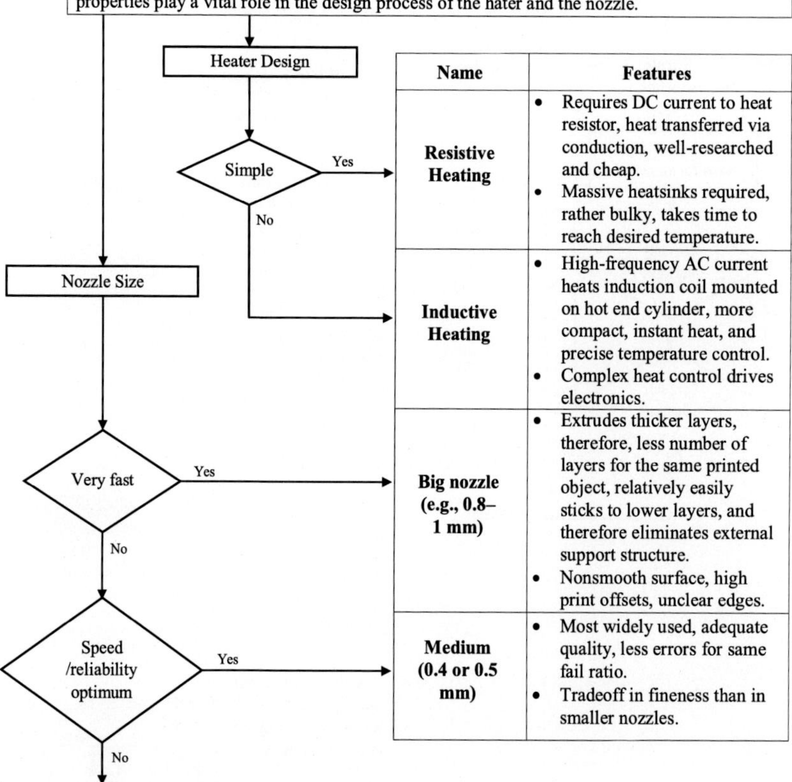

Name	Features
Resistive Heating	• Requires DC current to heat resistor, heat transferred via conduction, well-researched and cheap. • Massive heatsinks required, rather bulky, takes time to reach desired temperature.
Inductive Heating	• High-frequency AC current heats induction coil mounted on hot end cylinder, more compact, instant heat, and precise temperature control. • Complex heat control drives electronics.
Big nozzle (e.g., 0.8–1 mm)	• Extrudes thicker layers, therefore, less number of layers for the same printed object, relatively easily sticks to lower layers, and therefore eliminates external support structure. • Nonsmooth surface, high print offsets, unclear edges.
Medium (0.4 or 0.5 mm)	• Most widely used, adequate quality, less errors for same fail ratio. • Tradeoff in fineness than in smaller nozzles.

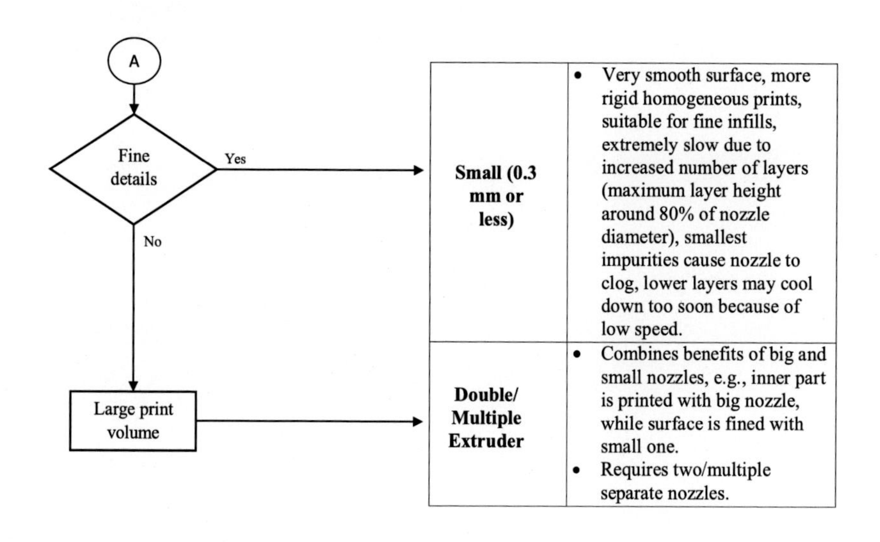

Name	Features
Small (0.3 mm or less)	• Very smooth surface, more rigid homogeneous prints, suitable for fine infills, extremely slow due to increased number of layers (maximum layer height around 80% of nozzle diameter), smallest impurities cause nozzle to clog, lower layers may cool down too soon because of low speed.
Double/ Multiple Extruder	• Combines benefits of big and small nozzles, e.g., inner part is printed with big nozzle, while surface is fined with small one. • Requires two/multiple separate nozzles.

Motion Types for 3D Printers

The differences of the motion for 3D printers comes from number of axis, movement type and platform movement.

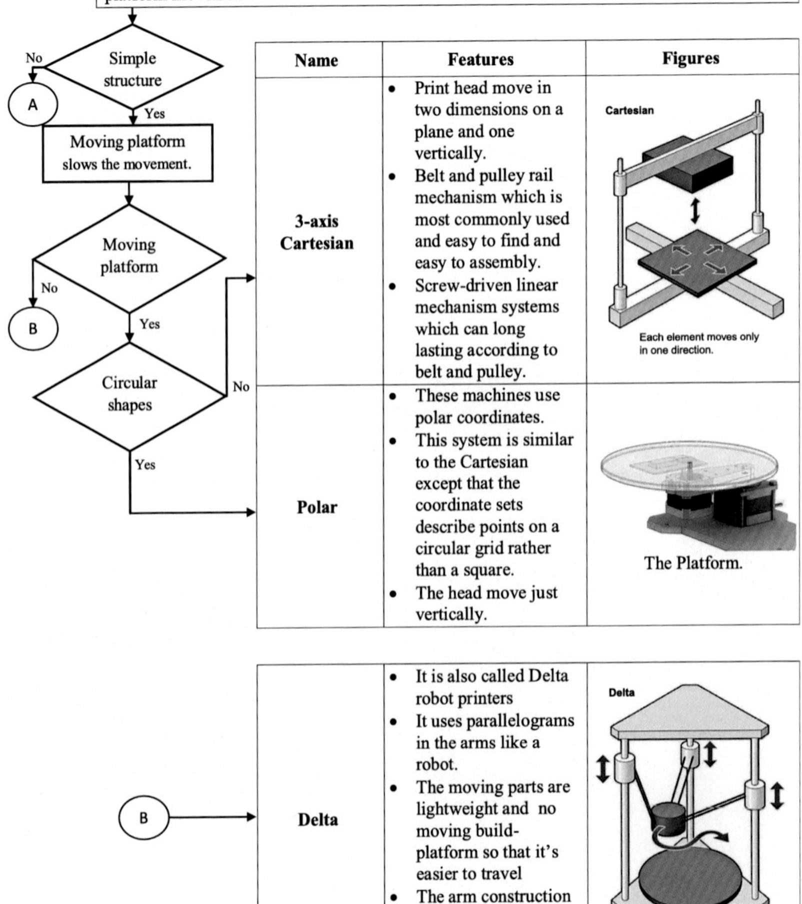

Name	Features	Figures
3-axis Cartesian	• Print head move in two dimensions on a plane and one vertically. • Belt and pulley rail mechanism which is most commonly used and easy to find and easy to assembly. • Screw-driven linear mechanism systems which can long lasting according to belt and pulley.	Cartesian Each element moves only in one direction.
Polar	• These machines use polar coordinates. • This system is similar to the Cartesian except that the coordinate sets describe points on a circular grid rather than a square. • The head move just vertically.	The Platform.

Delta	• It is also called Delta robot printers • It uses parallelograms in the arms like a robot. • The moving parts are lightweight and no moving build-platform so that it's easier to travel • The arm construction it must be a lot taller than your build volume.	Delta Printer head can move in any direction quickly.

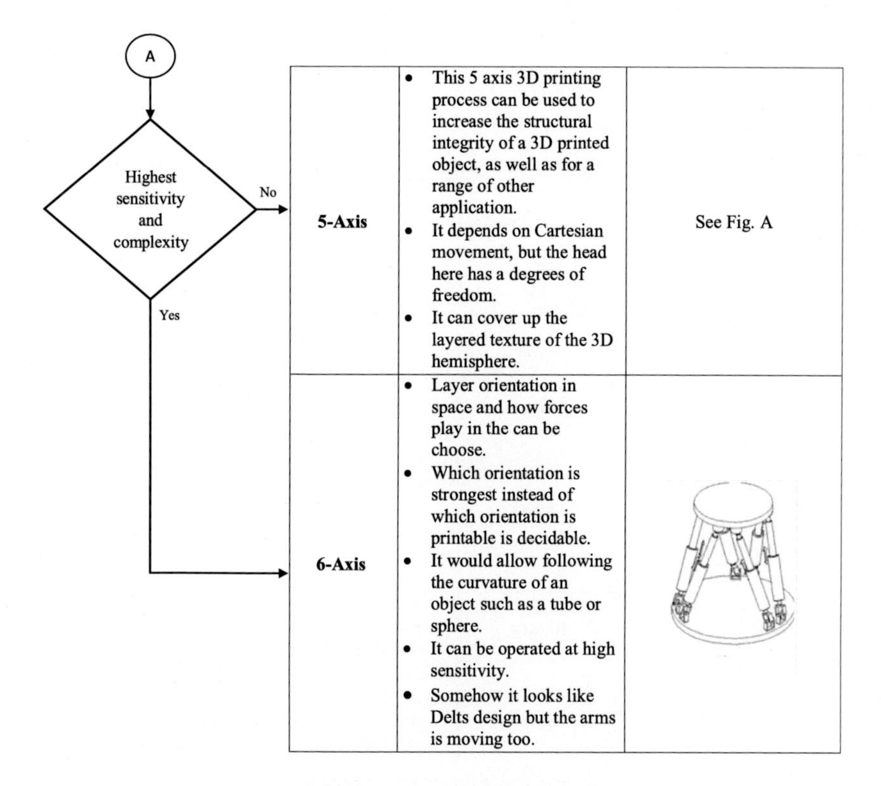

5-Axis	• This 5 axis 3D printing process can be used to increase the structural integrity of a 3D printed object, as well as for a range of other application. • It depends on Cartesian movement, but the head here has a degrees of freedom. • It can cover up the layered texture of the 3D hemisphere.	See Fig. A
6-Axis	• Layer orientation in space and how forces play in the can be choose. • Which orientation is strongest instead of which orientation is printable is decidable. • It would allow following the curvature of an object such as a tube or sphere. • It can be operated at high sensitivity. • Somehow it looks like Delts design but the arms is moving too.	

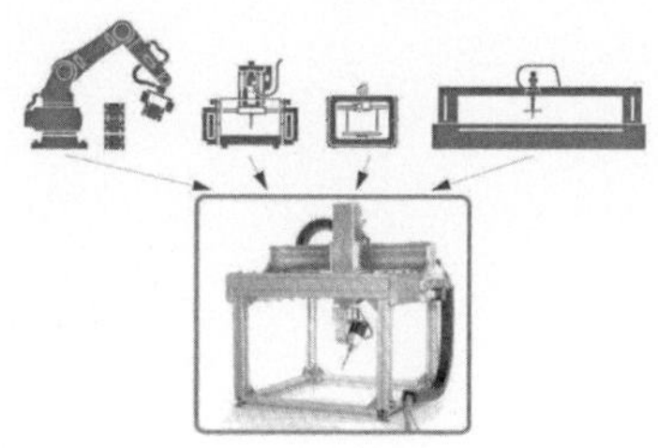

Fig.A: 5-Axis 3D printer structure

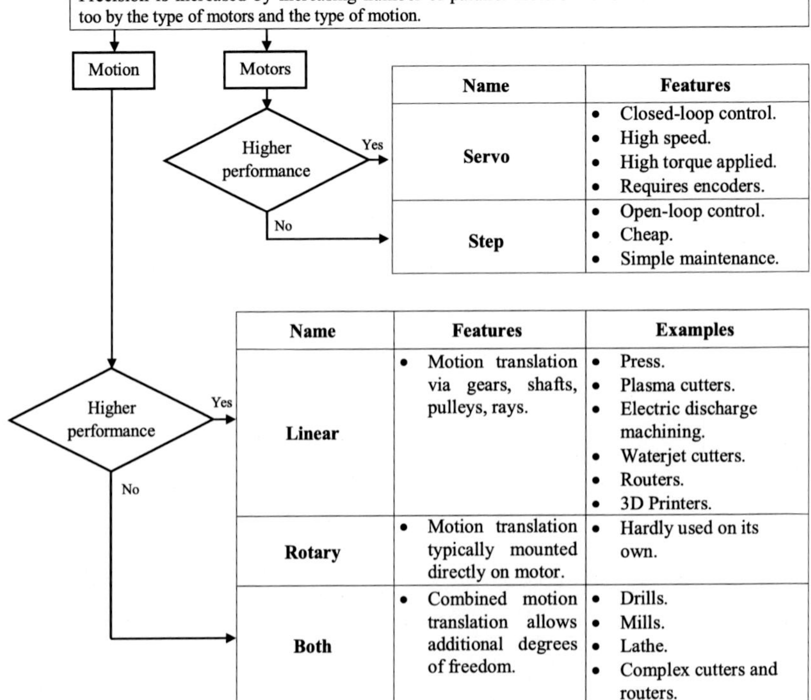

Name	Features
Servo	• Closed-loop control. • High speed. • High torque applied. • Requires encoders.
Step	• Open-loop control. • Cheap. • Simple maintenance.

Name	Features	Examples
Linear	• Motion translation via gears, shafts, pulleys, rays.	• Press. • Plasma cutters. • Electric discharge machining. • Waterjet cutters. • Routers. • 3D Printers.
Rotary	• Motion translation typically mounted directly on motor.	• Hardly used on its own.
Both	• Combined motion translation allows additional degrees of freedom.	• Drills. • Mills. • Lathe. • Complex cutters and routers.

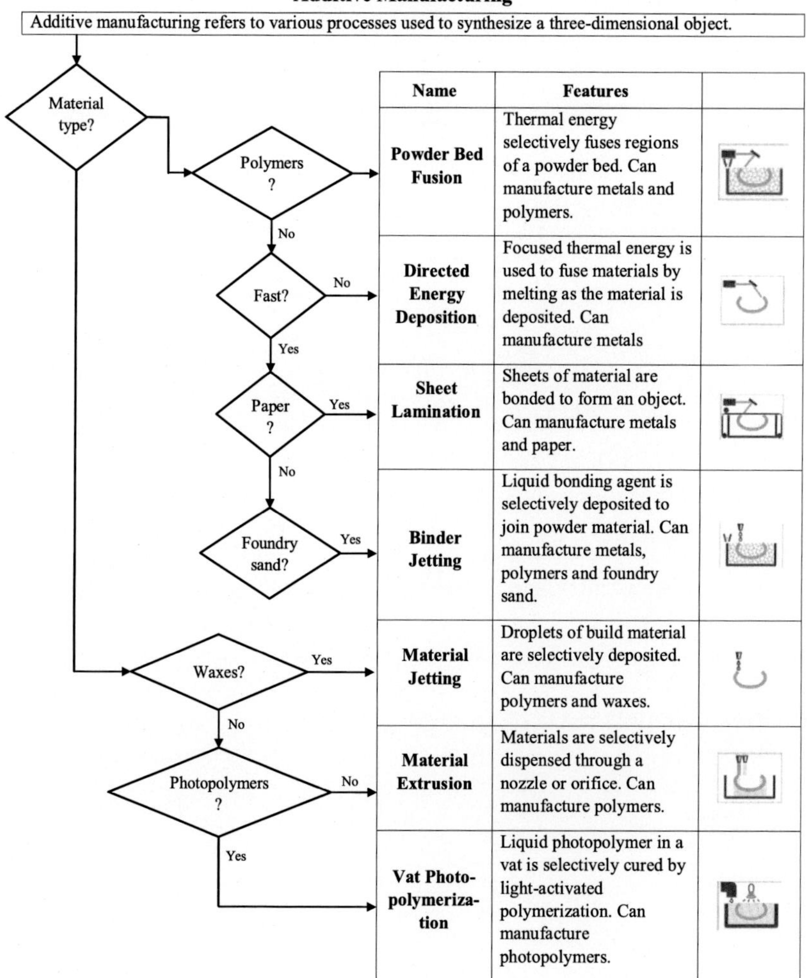

Name	Features	
Powder Bed Fusion	Thermal energy selectively fuses regions of a powder bed. Can manufacture metals and polymers.	
Directed Energy Deposition	Focused thermal energy is used to fuse materials by melting as the material is deposited. Can manufacture metals	
Sheet Lamination	Sheets of material are bonded to form an object. Can manufacture metals and paper.	
Binder Jetting	Liquid bonding agent is selectively deposited to join powder material. Can manufacture metals, polymers and foundry sand.	
Material Jetting	Droplets of build material are selectively deposited. Can manufacture polymers and waxes.	
Material Extrusion	Materials are selectively dispensed through a nozzle or orifice. Can manufacture polymers.	
Vat Photo-polymeriza-tion	Liquid photopolymer in a vat is selectively cured by light-activated polymerization. Can manufacture photopolymers.	

Chapter 7

Machine Elements

ABSTRACT

Machine elements are the basic parts of machines. Since these parts are manufactured in bulk, it is always cheaper to buy these parts than to manufacture them. Moreover, if possible, the designer should use machine elements to make servicing the machine easier in the future. Some of the functions inside a machine, such as transmission of power or reduction of speed, are a common challenge and can be easily solved with standard machine elements. The standardization of the machine elements makes them easier to use. The designer should learn both the functional difference and the standard of them. For example, gears, belts, and chains can be used for the transmission of power, but they have different advantages and disadvantages. Moreover, the designer should learn the standard of the interested part. For instance, the calculations can point to a M9 size bolt, but the standard sizes are M8 and M10. In this case, the M10 size bolt should be preferred to carry the force.

Mechatronic Components. https://doi.org/10.1016/B978-0-12-814126-7.00007-4

Mechanical Power Transmission Type

Power transmission is achieved by rotating machine part having cut teeth or cogs, which mesh with another toothed part to transmit torque, in most cases with teeth on the one gear being of identical shape.

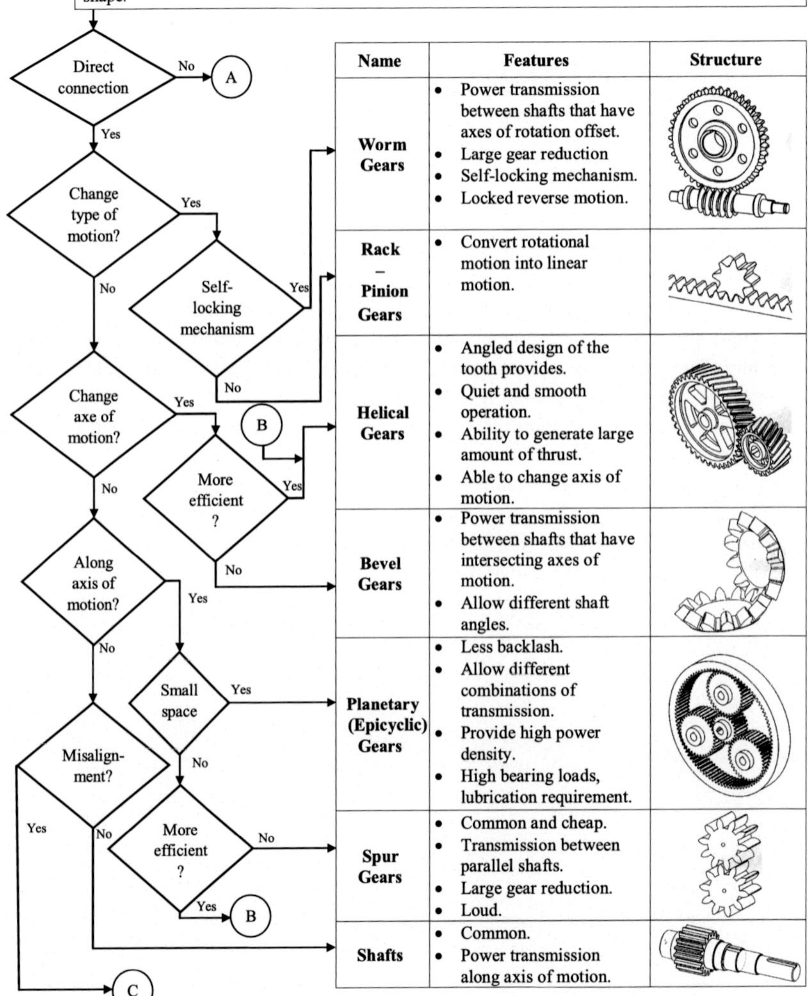

Name	Features	Structure
Worm Gears	• Power transmission between shafts that have axes of rotation offset. • Large gear reduction • Self-locking mechanism. • Locked reverse motion.	
Rack – Pinion Gears	• Convert rotational motion into linear motion.	
Helical Gears	• Angled design of the tooth provides. • Quiet and smooth operation. • Ability to generate large amount of thrust. • Able to change axis of motion.	
Bevel Gears	• Power transmission between shafts that have intersecting axes of motion. • Allow different shaft angles.	
Planetary (Epicyclic) Gears	• Less backlash. • Allow different combinations of transmission. • Provide high power density. • High bearing loads, lubrication requirement.	
Spur Gears	• Common and cheap. • Transmission between parallel shafts. • Large gear reduction. • Loud.	
Shafts	• Common. • Power transmission along axis of motion.	

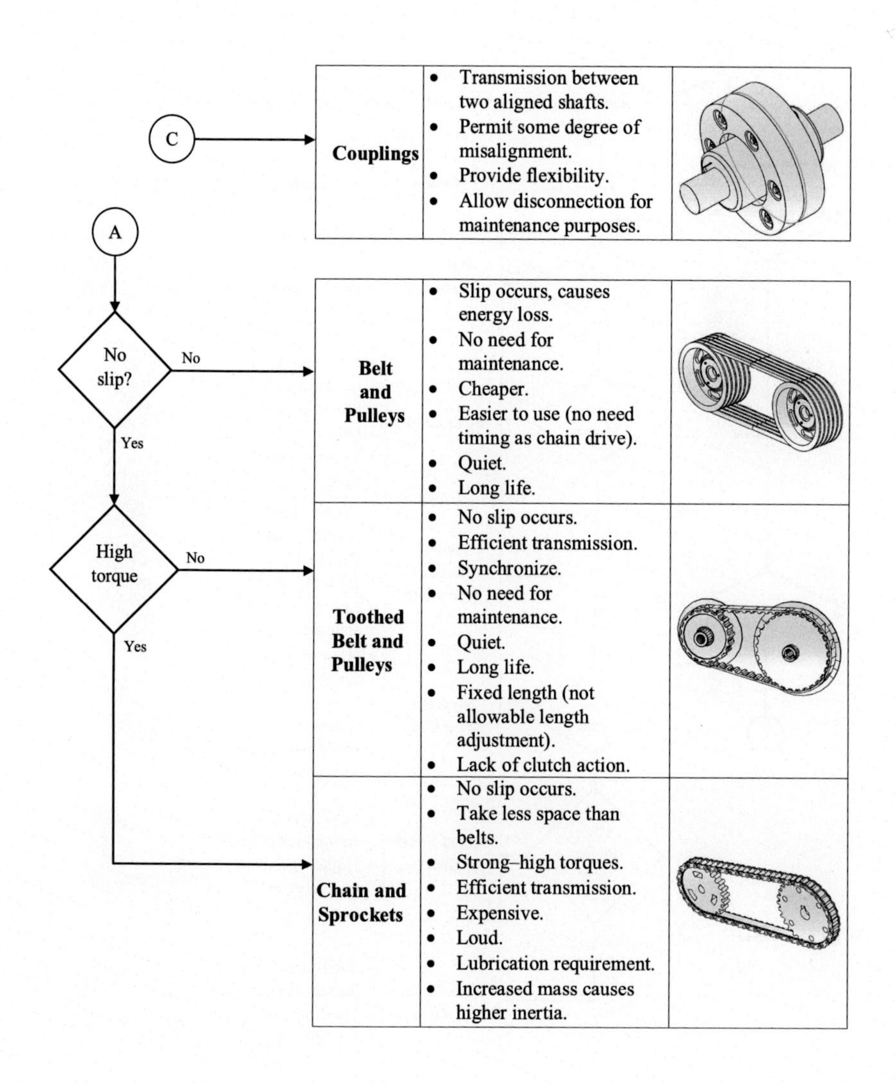

Couplings	• Transmission between two aligned shafts. • Permit some degree of misalignment. • Provide flexibility. • Allow disconnection for maintenance purposes.	
Belt and Pulleys	• Slip occurs, causes energy loss. • No need for maintenance. • Cheaper. • Easier to use (no need timing as chain drive). • Quiet. • Long life.	
Toothed Belt and Pulleys	• No slip occurs. • Efficient transmission. • Synchronize. • No need for maintenance. • Quiet. • Long life. • Fixed length (not allowable length adjustment). • Lack of clutch action.	
Chain and Sprockets	• No slip occurs. • Take less space than belts. • Strong–high torques. • Efficient transmission. • Expensive. • Loud. • Lubrication requirement. • Increased mass causes higher inertia.	

Gears

Gears are wheels with teeth. Gears mesh together and make things turn and are used to transfer motion or power from one moving part to another.

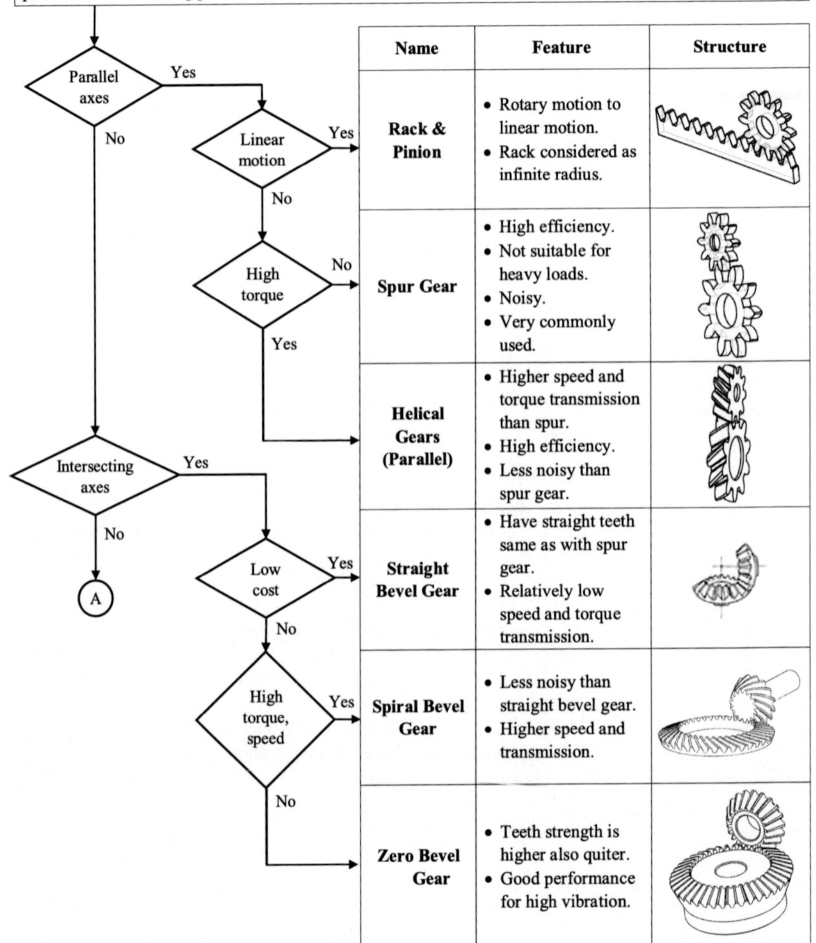

Name	Feature	Structure
Rack & Pinion	• Rotary motion to linear motion. • Rack considered as infinite radius.	
Spur Gear	• High efficiency. • Not suitable for heavy loads. • Noisy. • Very commonly used.	
Helical Gears (Parallel)	• Higher speed and torque transmission than spur. • High efficiency. • Less noisy than spur gear.	
Straight Bevel Gear	• Have straight teeth same as with spur gear. • Relatively low speed and torque transmission.	
Spiral Bevel Gear	• Less noisy than straight bevel gear. • Higher speed and transmission.	
Zero Bevel Gear	• Teeth strength is higher also quiter. • Good performance for high vibration.	

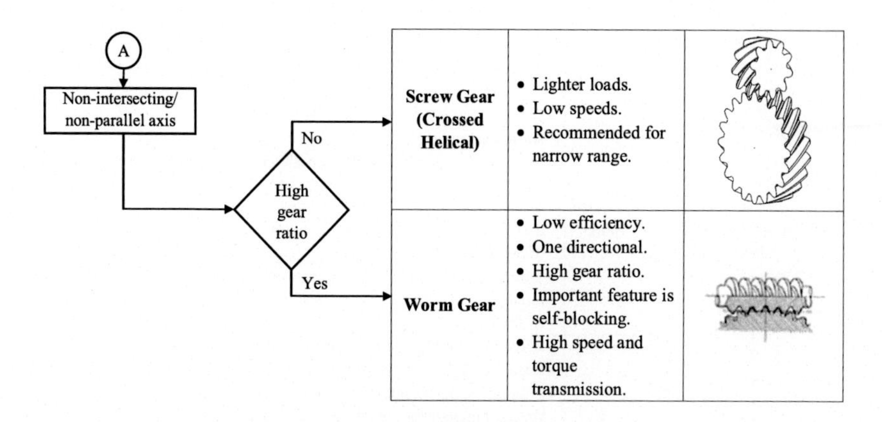

Screw Gear (Crossed Helical)	• Lighter loads. • Low speeds. • Recommended for narrow range.	
Worm Gear	• Low efficiency. • One directional. • High gear ratio. • Important feature is self-blocking. • High speed and torque transmission.	

Belts

A belt is a loop of flexible material used to link two or more rotating shafts mechanically. Belts may be used as a source of motion, to transmit power efficiently, or to track relative movement.

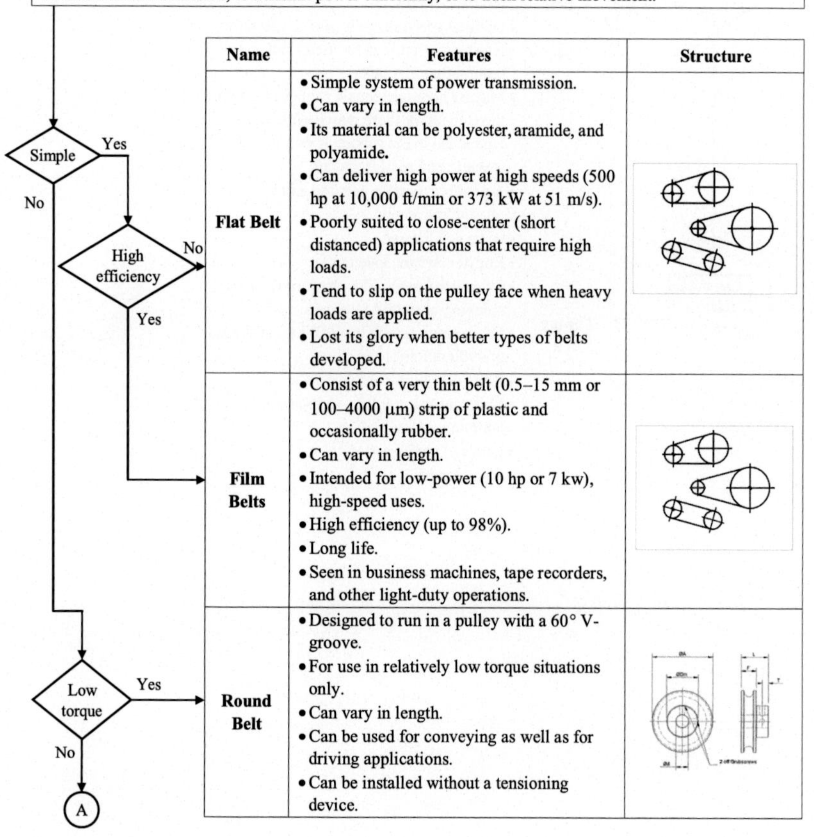

Name	Features	Structure
Flat Belt	• Simple system of power transmission. • Can vary in length. • Its material can be polyester, aramide, and polyamide. • Can deliver high power at high speeds (500 hp at 10,000 ft/min or 373 kW at 51 m/s). • Poorly suited to close-center (short distanced) applications that require high loads. • Tend to slip on the pulley face when heavy loads are applied. • Lost its glory when better types of belts developed.	
Film Belts	• Consist of a very thin belt (0.5–15 mm or 100–4000 μm) strip of plastic and occasionally rubber. • Can vary in length. • Intended for low-power (10 hp or 7 kw), high-speed uses. • High efficiency (up to 98%). • Long life. • Seen in business machines, tape recorders, and other light-duty operations.	
Round Belt	• Designed to run in a pulley with a 60° V-groove. • For use in relatively low torque situations only. • Can vary in length. • Can be used for conveying as well as for driving applications. • Can be installed without a tensioning device.	

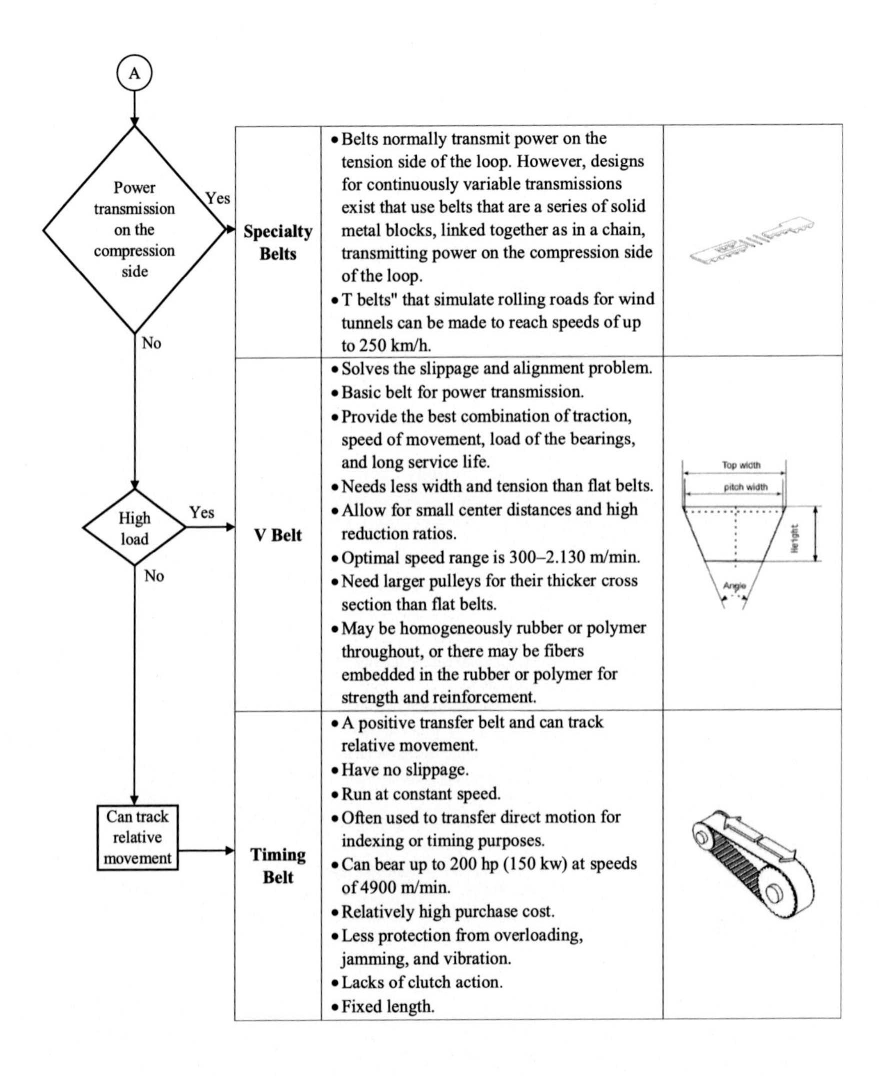

Specialty Belts	• Belts normally transmit power on the tension side of the loop. However, designs for continuously variable transmissions exist that use belts that are a series of solid metal blocks, linked together as in a chain, transmitting power on the compression side of the loop. • T belts" that simulate rolling roads for wind tunnels can be made to reach speeds of up to 250 km/h.	
V Belt	• Solves the slippage and alignment problem. • Basic belt for power transmission. • Provide the best combination of traction, speed of movement, load of the bearings, and long service life. • Needs less width and tension than flat belts. • Allow for small center distances and high reduction ratios. • Optimal speed range is 300–2.130 m/min. • Need larger pulleys for their thicker cross section than flat belts. • May be homogeneously rubber or polymer throughout, or there may be fibers embedded in the rubber or polymer for strength and reinforcement.	
Timing Belt	• A positive transfer belt and can track relative movement. • Have no slippage. • Run at constant speed. • Often used to transfer direct motion for indexing or timing purposes. • Can bear up to 200 hp (150 kw) at speeds of 4900 m/min. • Relatively high purchase cost. • Less protection from overloading, jamming, and vibration. • Lacks of clutch action. • Fixed length.	

Mechanical Keys

According to the task of the joint, it is distinguished between fastening and tightening key joints. Keys, used in fastening key joints, are distinguished by their position to the longitudinal axis of the machine parts to be connected.

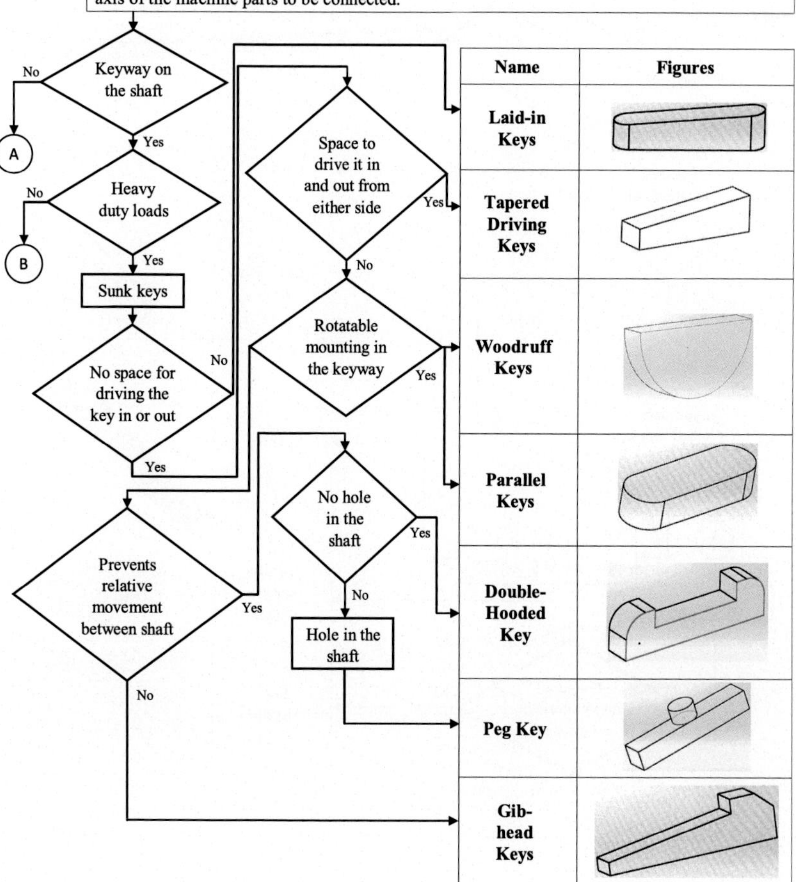

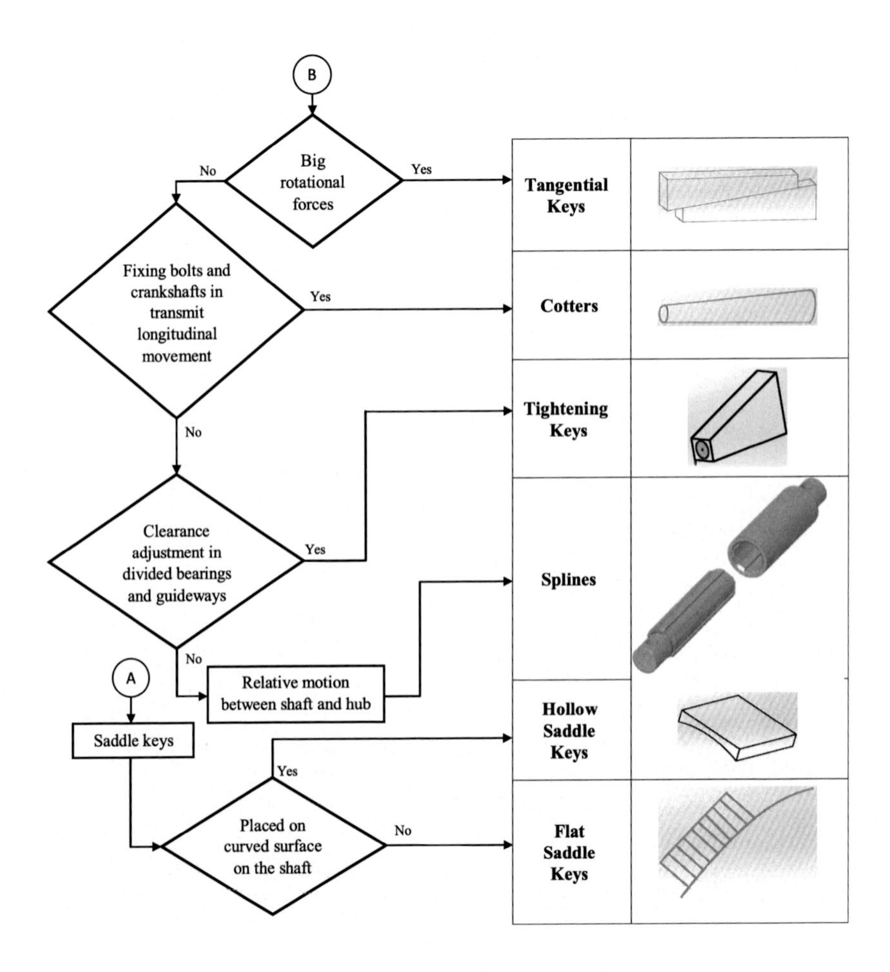
B
Big rotational forces
No
Yes
Tangential Keys
Fixing bolts and crankshafts in transmit longitudinal movement
Yes
Cotters
No
Tightening Keys
Clearance adjustment in divided bearings and guideways
Yes
Splines
No
Relative motion between shaft and hub
A
Saddle keys
Hollow Saddle Keys
Yes
Placed on curved surface on the shaft
No
Flat Saddle Keys

Brakes and Clutches

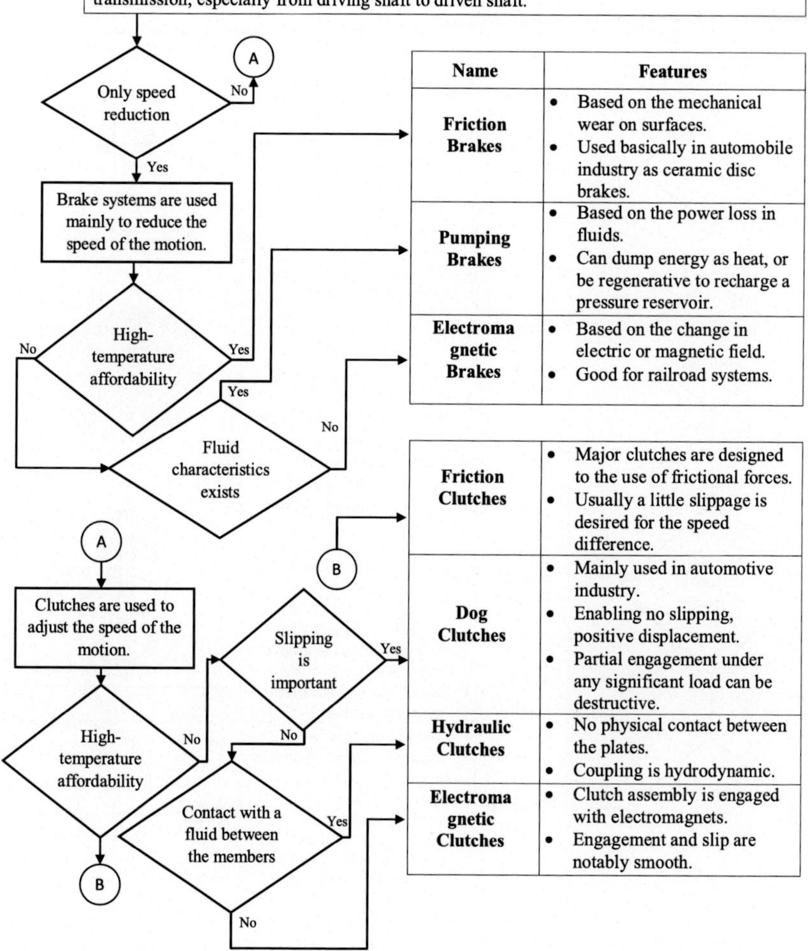

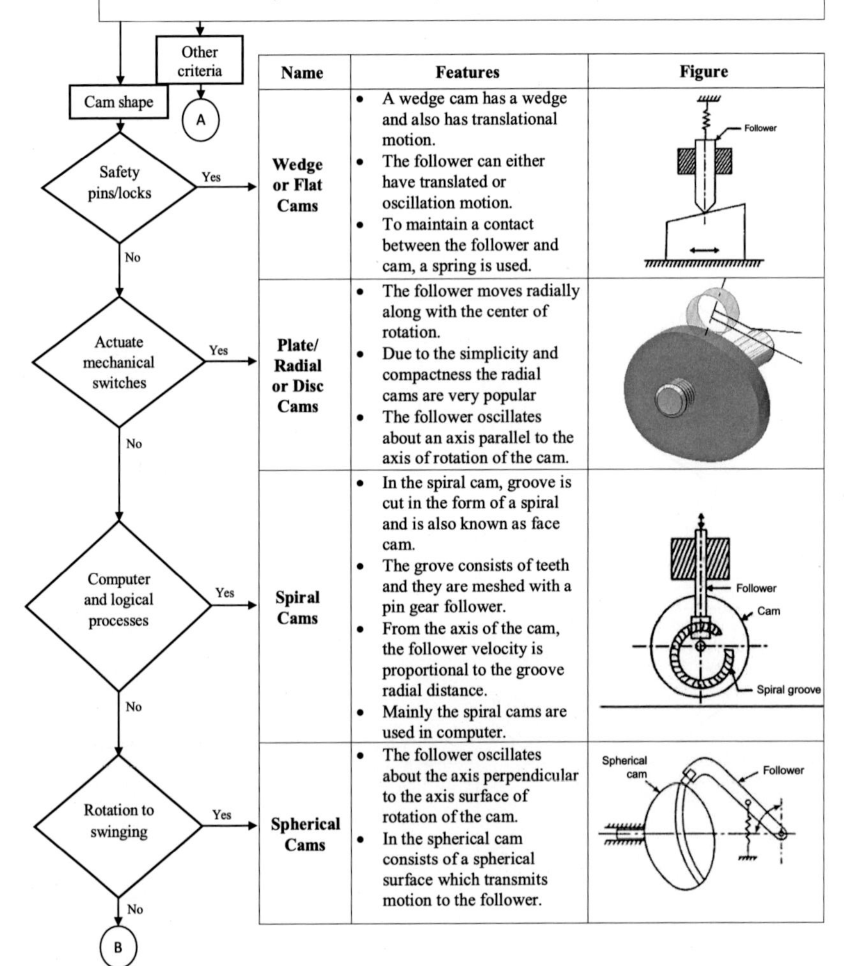

Name	Features	Figure
Wedge or Flat Cams	• A wedge cam has a wedge and also has translational motion. • The follower can either have translated or oscillation motion. • To maintain a contact between the follower and cam, a spring is used.	
Plate/ Radial or Disc Cams	• The follower moves radially along with the center of rotation. • Due to the simplicity and compactness the radial cams are very popular • The follower oscillates about an axis parallel to the axis of rotation of the cam.	
Spiral Cams	• In the spiral cam, groove is cut in the form of a spiral and is also known as face cam. • The grove consists of teeth and they are meshed with a pin gear follower. • From the axis of the cam, the follower velocity is proportional to the groove radial distance. • Mainly the spiral cams are used in computer.	
Spherical Cams	• The follower oscillates about the axis perpendicular to the axis surface of rotation of the cam. • In the spherical cam consists of a spherical surface which transmits motion to the follower.	

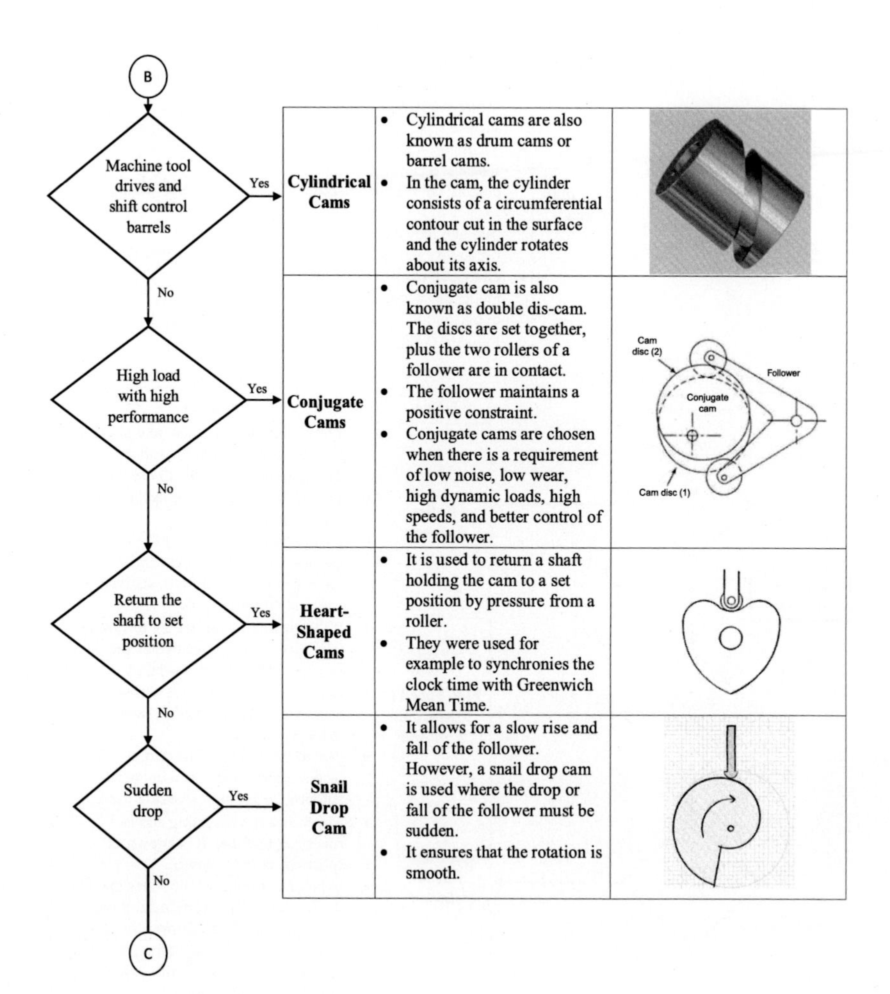

Cylindrical Cams	• Cylindrical cams are also known as drum cams or barrel cams. • In the cam, the cylinder consists of a circumferential contour cut in the surface and the cylinder rotates about its axis.	
Conjugate Cams	• Conjugate cam is also known as double dis-cam. The discs are set together, plus the two rollers of a follower are in contact. • The follower maintains a positive constraint. • Conjugate cams are chosen when there is a requirement of low noise, low wear, high dynamic loads, high speeds, and better control of the follower.	
Heart-Shaped Cams	• It is used to return a shaft holding the cam to a set position by pressure from a roller. • They were used for example to synchronies the clock time with Greenwich Mean Time.	
Snail Drop Cam	• It allows for a slow rise and fall of the follower. However, a snail drop cam is used where the drop or fall of the follower must be sudden. • It ensures that the rotation is smooth.	

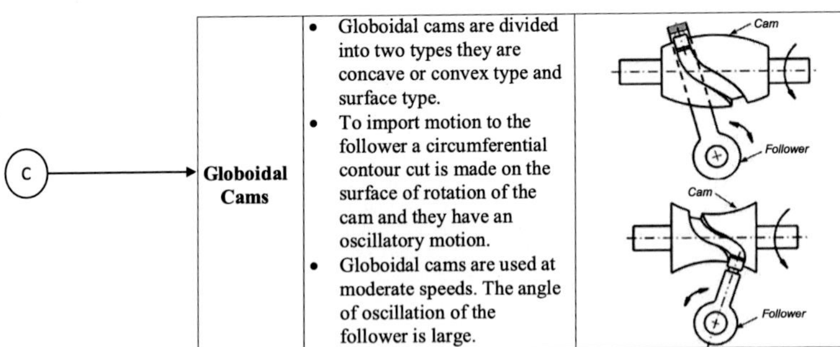

Globoidal Cams	• Globoidal cams are divided into two types they are concave or convex type and surface type. • To import motion to the follower a circumferential contour cut is made on the surface of rotation of the cam and they have an oscillatory motion. • Globoidal cams are used at moderate speeds. The angle of oscillation of the follower is large.	

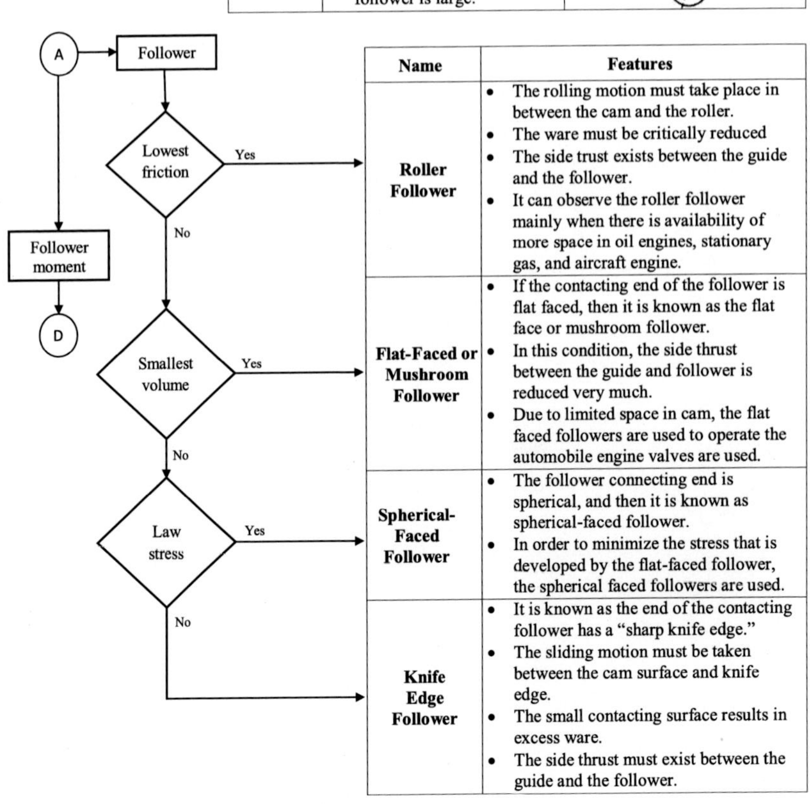

Name	Features
Roller Follower	• The rolling motion must take place in between the cam and the roller. • The ware must be critically reduced • The side trust exists between the guide and the follower. • It can observe the roller follower mainly when there is availability of more space in oil engines, stationary gas, and aircraft engine.
Flat-Faced or Mushroom Follower	• If the contacting end of the follower is flat faced, then it is known as the flat face or mushroom follower. • In this condition, the side thrust between the guide and follower is reduced very much. • Due to limited space in cam, the flat faced followers are used to operate the automobile engine valves are used.
Spherical-Faced Follower	• The follower connecting end is spherical, and then it is known as spherical-faced follower. • In order to minimize the stress that is developed by the flat-faced follower, the spherical faced followers are used.
Knife Edge Follower	• It is known as the end of the contacting follower has a "sharp knife edge." • The sliding motion must be taken between the cam surface and knife edge. • The small contacting surface results in excess ware. • The side thrust must exist between the guide and the follower.

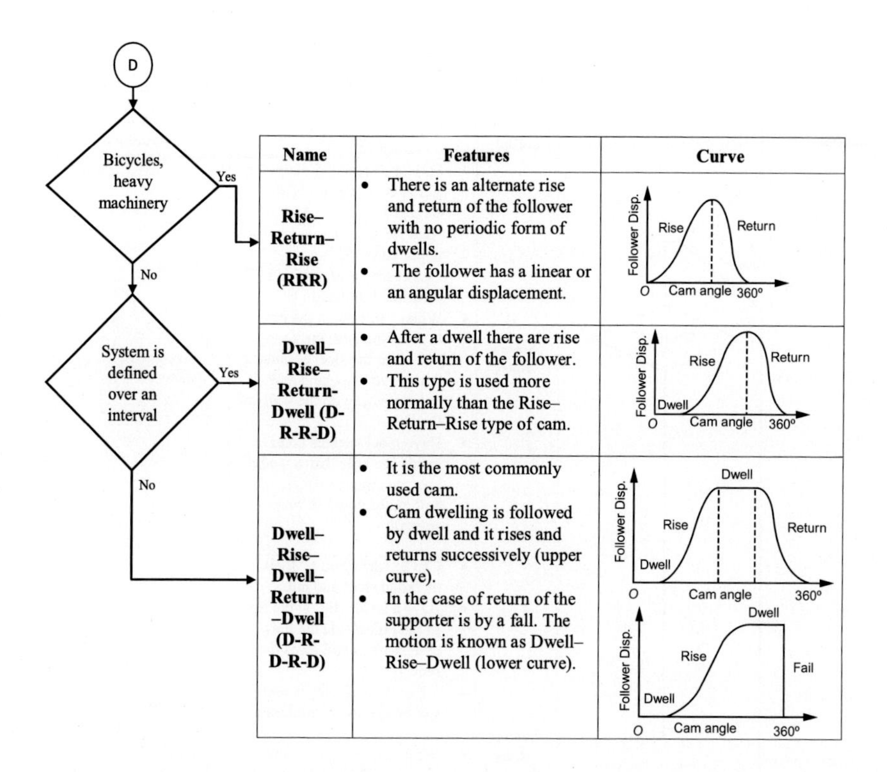

Name	Features	Curve
Rise–Return–Rise (RRR)	• There is an alternate rise and return of the follower with no periodic form of dwells. • The follower has a linear or an angular displacement.	Follower Disp.; Rise; Return; O; Cam angle; 360°
Dwell–Rise–Return-Dwell (D-R-R-D)	• After a dwell there are rise and return of the follower. • This type is used more normally than the Rise–Return–Rise type of cam.	Follower Disp.; Dwell; Rise; Return; O; Cam angle; 360°
Dwell–Rise–Dwell–Return–Dwell (D-R-D-R-D)	• It is the most commonly used cam. • Cam dwelling is followed by dwell and it rises and returns successively (upper curve). • In the case of return of the supporter is by a fall. The motion is known as Dwell–Rise–Dwell (lower curve).	Follower Disp.; Dwell; Rise; Dwell; Return; O; Cam angle; 360° Follower Disp.; Dwell; Rise; Dwell; Fail; O; Cam angle; 360°

Bolts

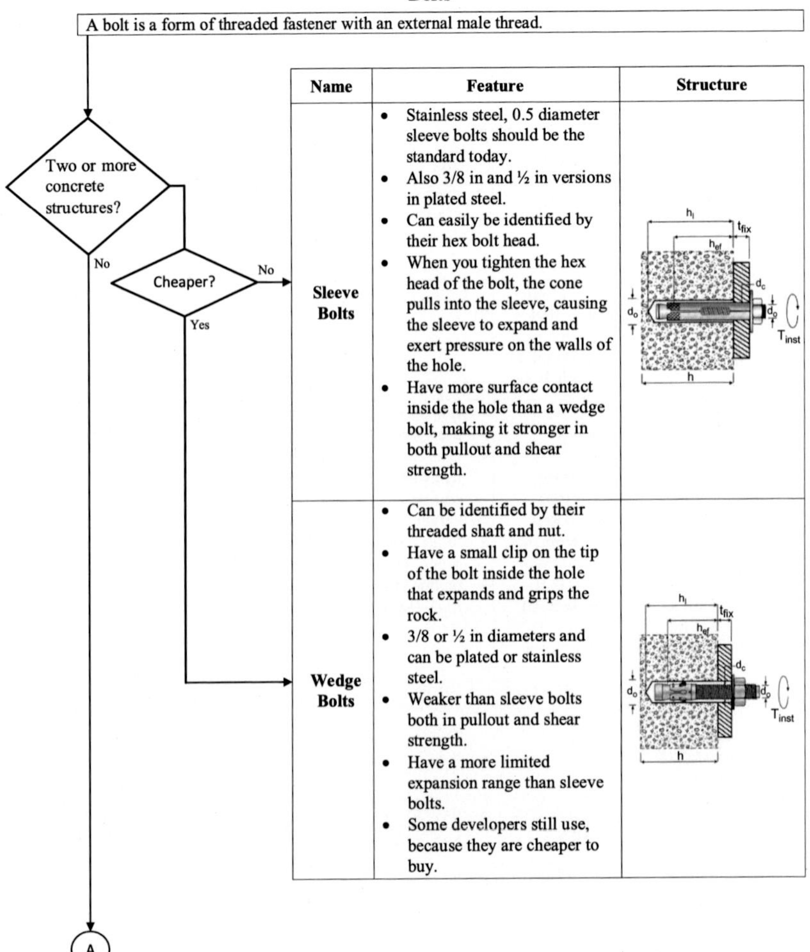

Name	Feature	Structure
Sleeve Bolts	• Stainless steel, 0.5 diameter sleeve bolts should be the standard today. • Also 3/8 in and ½ in versions in plated steel. • Can easily be identified by their hex bolt head. • When you tighten the hex head of the bolt, the cone pulls into the sleeve, causing the sleeve to expand and exert pressure on the walls of the hole. • Have more surface contact inside the hole than a wedge bolt, making it stronger in both pullout and shear strength.	h_l, t_{fix}, h_{ef}, d_c, d_o, d_p, T_{inst}, h
Wedge Bolts	• Can be identified by their threaded shaft and nut. • Have a small clip on the tip of the bolt inside the hole that expands and grips the rock. • 3/8 or ½ in diameters and can be plated or stainless steel. • Weaker than sleeve bolts both in pullout and shear strength. • Have a more limited expansion range than sleeve bolts. • Some developers still use, because they are cheaper to buy.	h_l, t_{fix}, h_{ef}, d_c, d_o, d_p, T_{inst}, h

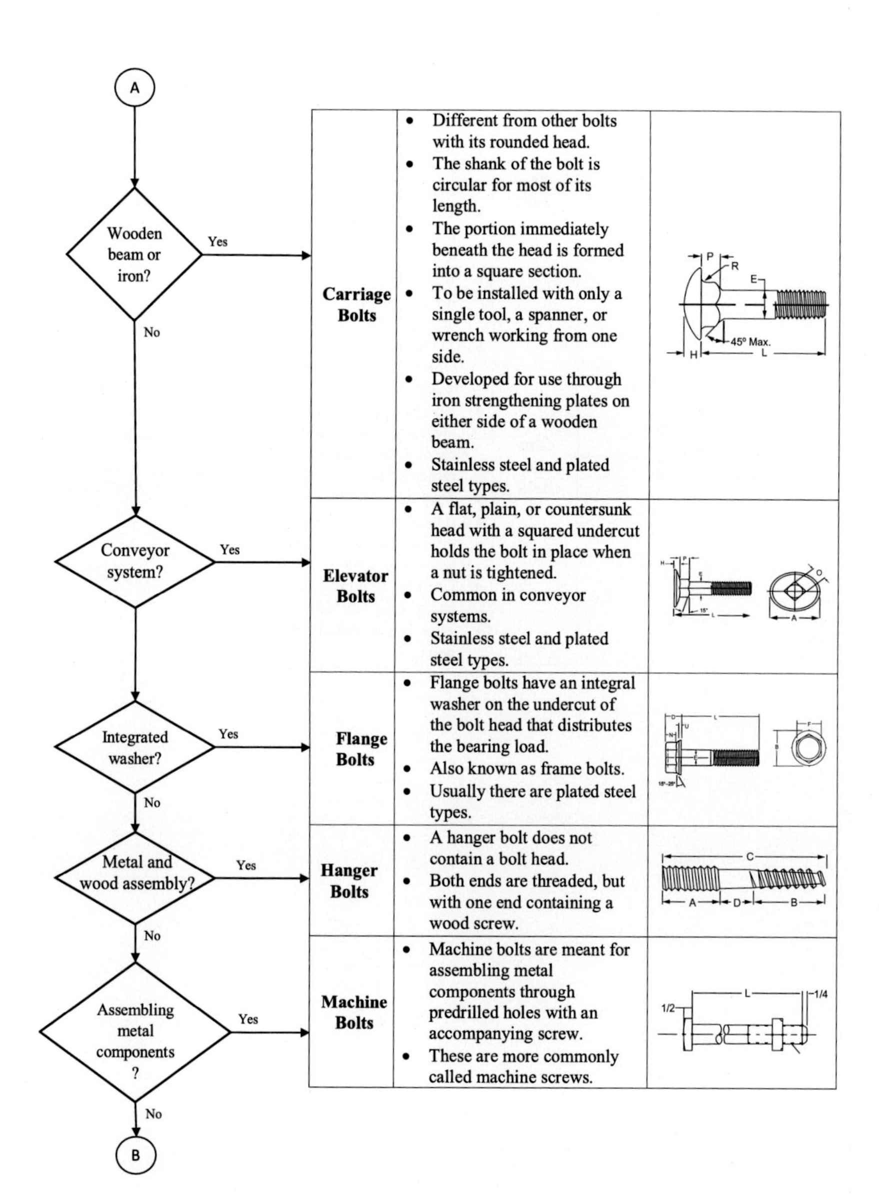

Carriage Bolts	• Different from other bolts with its rounded head. • The shank of the bolt is circular for most of its length. • The portion immediately beneath the head is formed into a square section. • To be installed with only a single tool, a spanner, or wrench working from one side. • Developed for use through iron strengthening plates on either side of a wooden beam. • Stainless steel and plated steel types.	
Elevator Bolts	• A flat, plain, or countersunk head with a squared undercut holds the bolt in place when a nut is tightened. • Common in conveyor systems. • Stainless steel and plated steel types.	
Flange Bolts	• Flange bolts have an integral washer on the undercut of the bolt head that distributes the bearing load. • Also known as frame bolts. • Usually there are plated steel types.	
Hanger Bolts	• A hanger bolt does not contain a bolt head. • Both ends are threaded, but with one end containing a wood screw.	
Machine Bolts	• Machine bolts are meant for assembling metal components through predrilled holes with an accompanying screw. • These are more commonly called machine screws.	

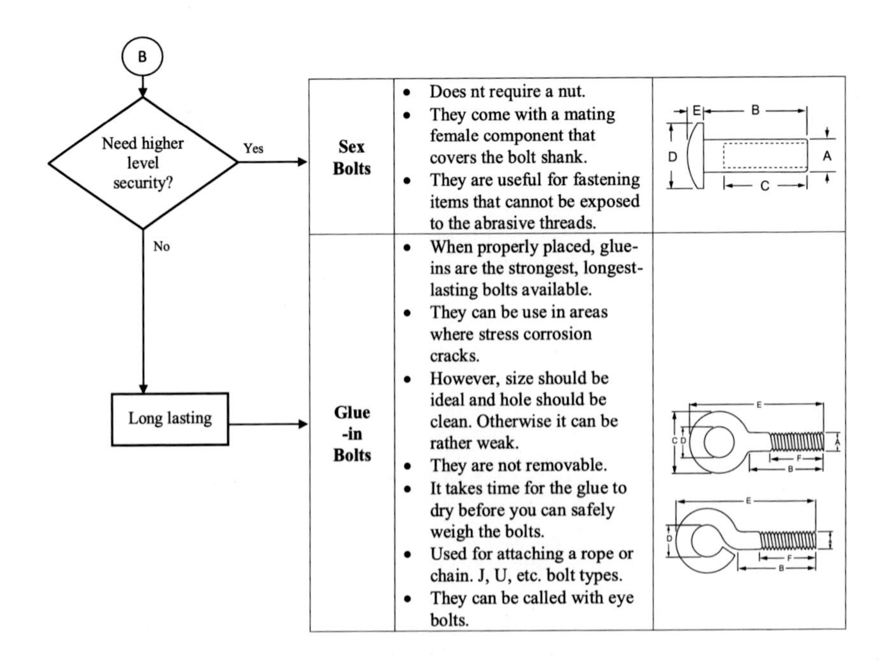

Sex Bolts	• Does nt require a nut. • They come with a mating female component that covers the bolt shank. • They are useful for fastening items that cannot be exposed to the abrasive threads.	
Glue -in Bolts	• When properly placed, glue-ins are the strongest, longest-lasting bolts available. • They can be use in areas where stress corrosion cracks. • However, size should be ideal and hole should be clean. Otherwise it can be rather weak. • They are not removable. • It takes time for the glue to dry before you can safely weigh the bolts. • Used for attaching a rope or chain. J, U, etc. bolt types. • They can be called with eye bolts.	

Chapter 8

Design and Analysis Programs

ABSTRACT

Computers in general have a very important effect on the design and analysis of mechatronic systems. For mechanical parts, computer-aided design programs improve the design process and for the electronics, circuit drawing programs make the design process very effective. With the help of the mechanical drawing programs, the mechanical parts can be created in the computer environment and for the manufacturing the CNC machines codes can be generated. By using 3D versions, the assembly drawings can be prepared. The circuit drawing programs can do a similar task for the electronic parts of the project. Especially some programs can recall parts from library files, making it very easy for the designer. Even some of the software can be incorporated with the software so that the software can also be controlled in a virtual environment.

Finite Element Analysis programs can run most of the mechanical failure mode calculations, making the part optimization easy for the designer.

Mechatronic Components. https://doi.org/10.1016/B978-0-12-814126-7.00008-6

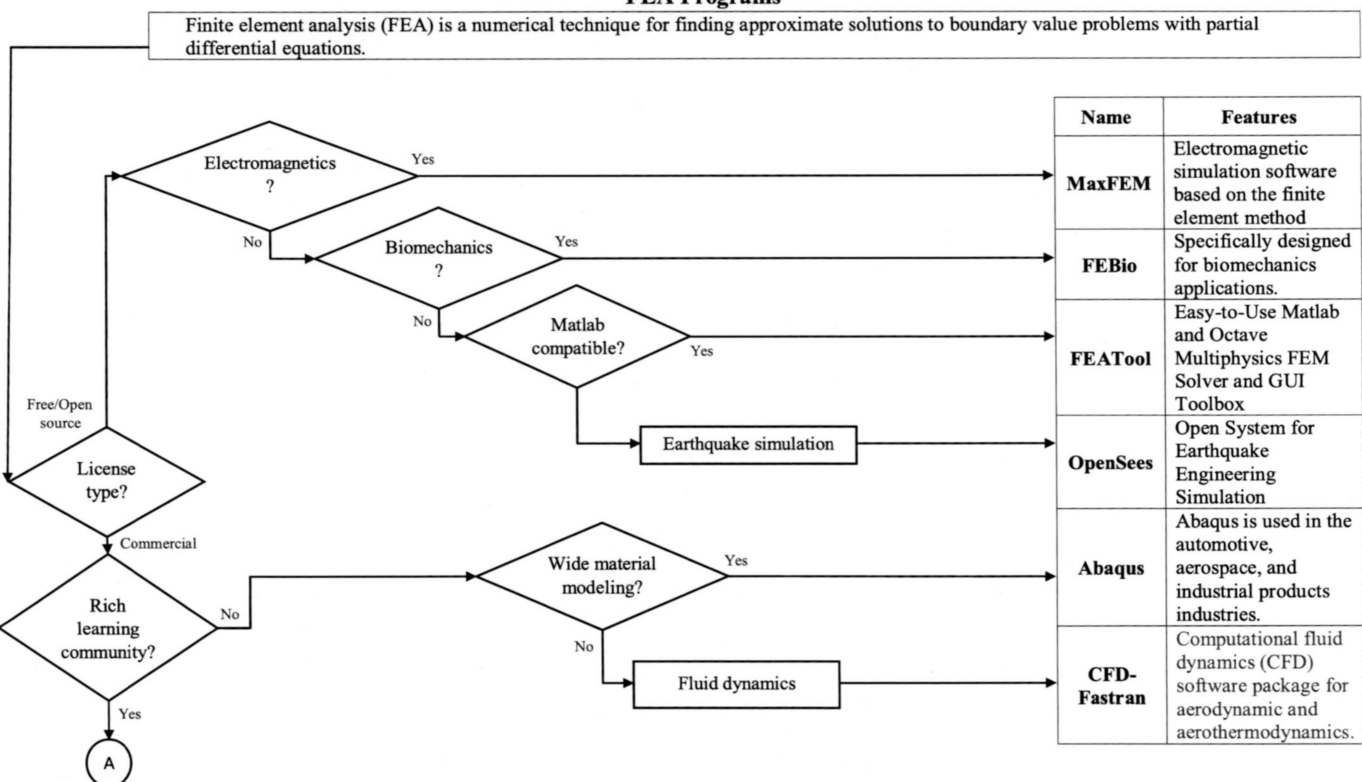

Name	Features
MaxFEM	Electromagnetic simulation software based on the finite element method
FEBio	Specifically designed for biomechanics applications.
FEATool	Easy-to-Use Matlab and Octave Multiphysics FEM Solver and GUI Toolbox
OpenSees	Open System for Earthquake Engineering Simulation
Abaqus	Abaqus is used in the automotive, aerospace, and industrial products industries.
CFD-Fastran	Computational fluid dynamics (CFD) software package for aerodynamic and aerothermodynamics.

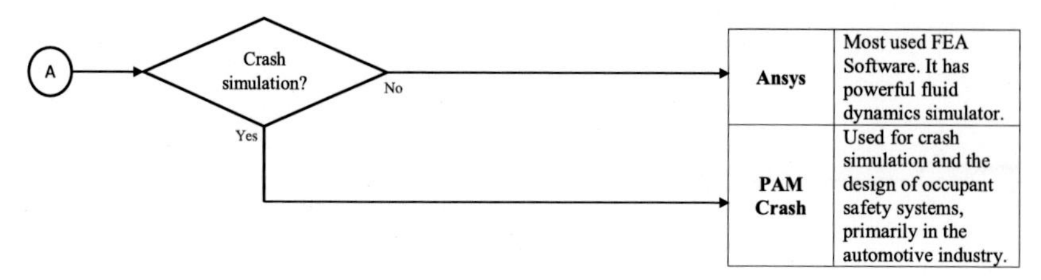

Drawing Programs

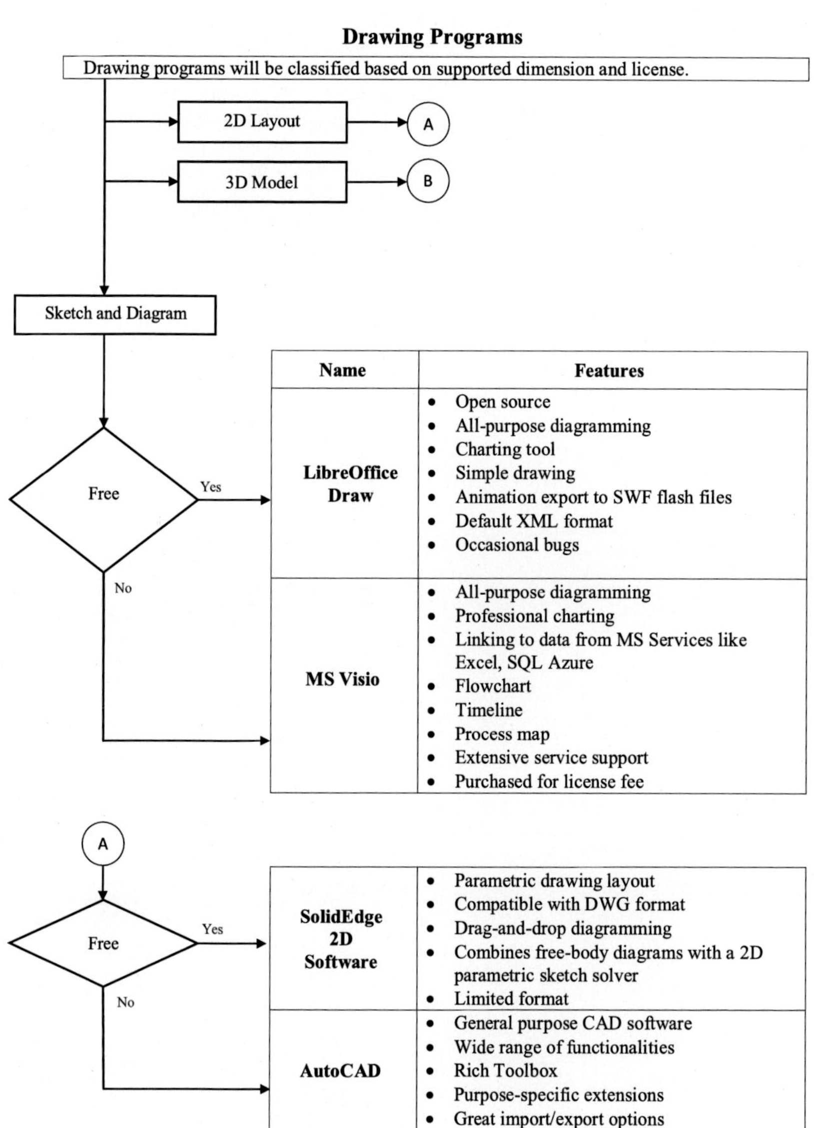

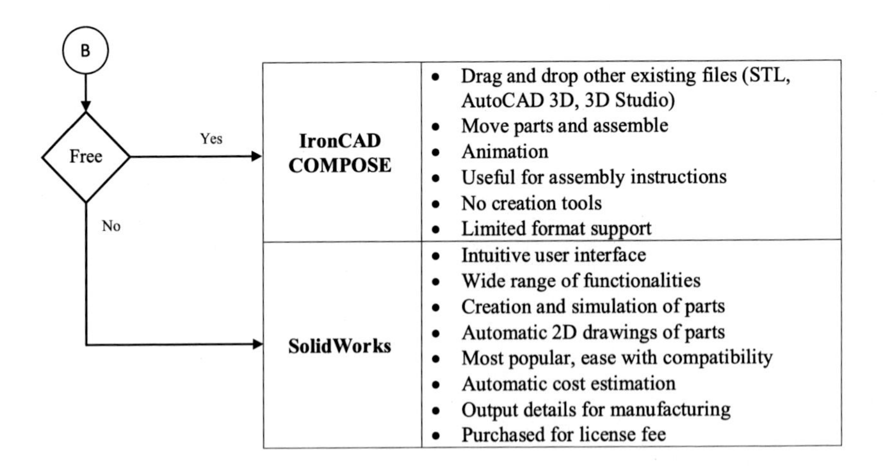
B
Free
Yes
No
IronCAD COMPOSE
Drag and drop other existing files (STL, AutoCAD 3D, 3D Studio)
Move parts and assemble
Animation
Useful for assembly instructions
No creation tools
Limited format support
SolidWorks
Intuitive user interface
Wide range of functionalities
Creation and simulation of parts
Automatic 2D drawings of parts
Most popular, ease with compatibility
Automatic cost estimation
Output details for manufacturing
Purchased for license fee

Meshing

The partial differential equations that govern a physical phenomenon (fluid flow, heat transfer, stress distribution etc.) are not usually amenable to analytical solutions, except for very simple cases. Therefore, in order to analyze a physical phenomenon, corresponding physical domains are split into smaller subdomains (cells or elements). The governing equations are then discretized and solved inside each of these subdomains and then put together to get a complete solution of the phenomenon inside the domain.

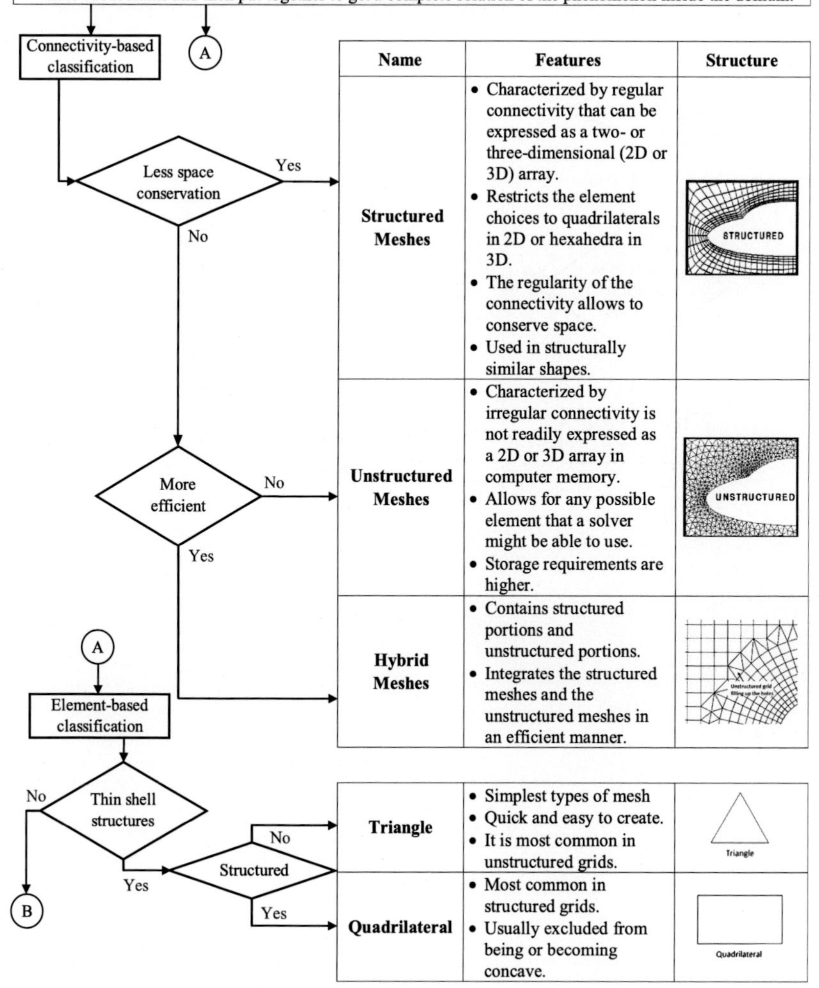

Name	Features	Structure
Structured Meshes	• Characterized by regular connectivity that can be expressed as a two- or three-dimensional (2D or 3D) array. • Restricts the element choices to quadrilaterals in 2D or hexahedra in 3D. • The regularity of the connectivity allows to conserve space. • Used in structurally similar shapes.	
Unstructured Meshes	• Characterized by irregular connectivity is not readily expressed as a 2D or 3D array in computer memory. • Allows for any possible element that a solver might be able to use. • Storage requirements are higher.	
Hybrid Meshes	• Contains structured portions and unstructured portions. • Integrates the structured meshes and the unstructured meshes in an efficient manner.	

Triangle	• Simplest types of mesh • Quick and easy to create. • It is most common in unstructured grids.	
Quadrilateral	• Most common in structured grids. • Usually excluded from being or becoming concave.	

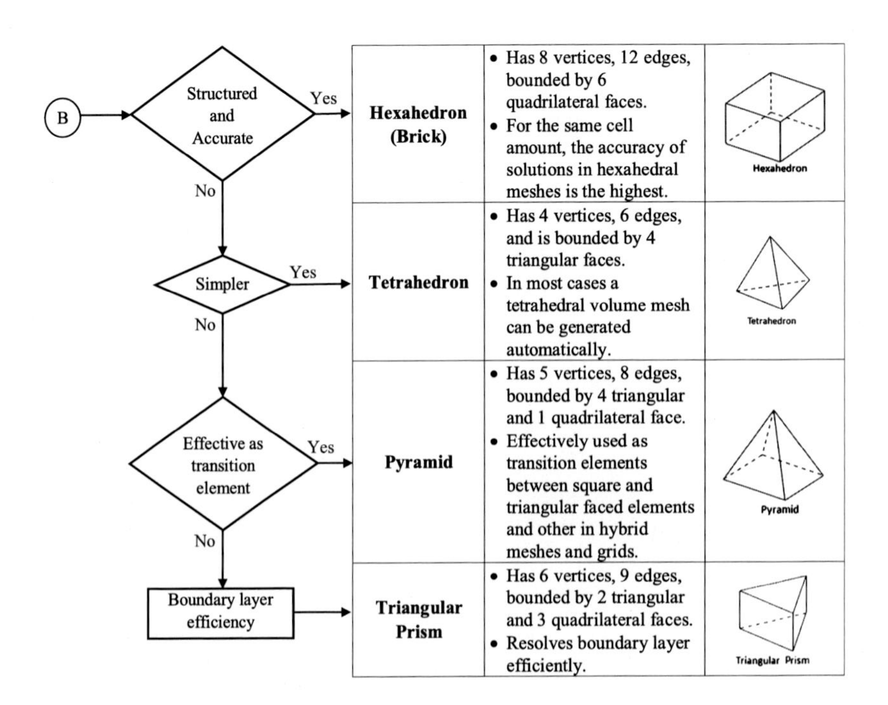
B
Structured and Accurate
Yes
No
Simpler
Yes
No
Effective as transition element
Yes
No
Boundary layer efficiency
Hexahedron (Brick)
• Has 8 vertices, 12 edges, bounded by 6 quadrilateral faces.
• For the same cell amount, the accuracy of solutions in hexahedral meshes is the highest.
Hexahedron
Tetrahedron
• Has 4 vertices, 6 edges, and is bounded by 4 triangular faces.
• In most cases a tetrahedral volume mesh can be generated automatically.
Tetrahedron
Pyramid
• Has 5 vertices, 8 edges, bounded by 4 triangular and 1 quadrilateral face.
• Effectively used as transition elements between square and triangular faced elements and other in hybrid meshes and grids.
Pyramid
Triangular Prism
• Has 6 vertices, 9 edges, bounded by 2 triangular and 3 quadrilateral faces.
• Resolves boundary layer efficiently.
Triangular Prism

Circuit Drawing Programs

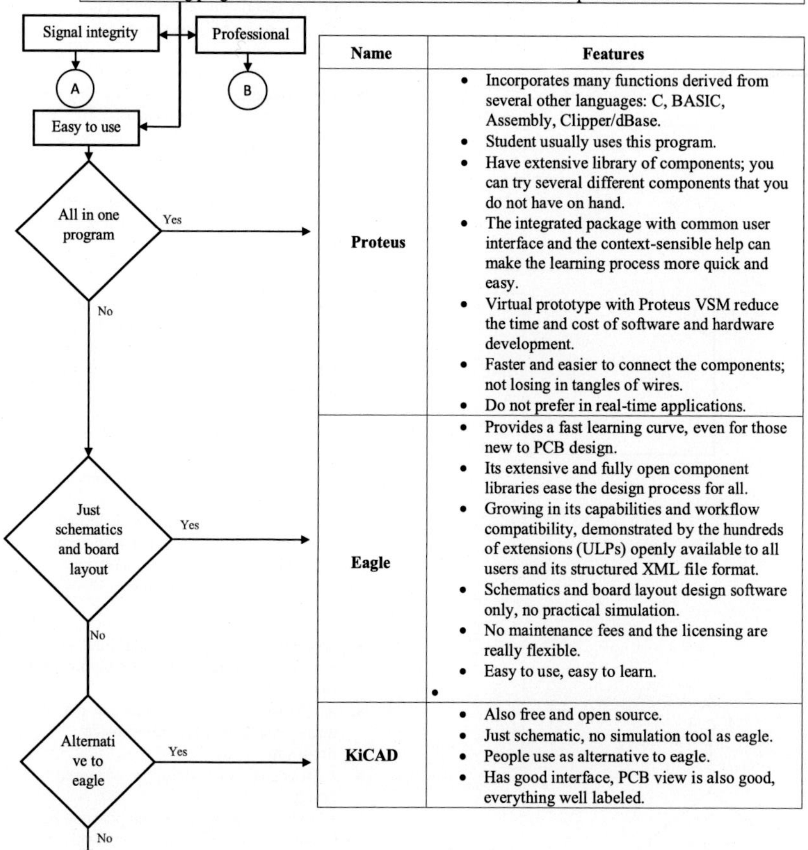

Name	Features
Proteus	• Incorporates many functions derived from several other languages: C, BASIC, Assembly, Clipper/dBase. • Student usually uses this program. • Have extensive library of components; you can try several different components that you do not have on hand. • The integrated package with common user interface and the context-sensible help can make the learning process more quick and easy. • Virtual prototype with Proteus VSM reduce the time and cost of software and hardware development. • Faster and easier to connect the components; not losing in tangles of wires. • Do not prefer in real-time applications.
Eagle	• Provides a fast learning curve, even for those new to PCB design. • Its extensive and fully open component libraries ease the design process for all. • Growing in its capabilities and workflow compatibility, demonstrated by the hundreds of extensions (ULPs) openly available to all users and its structured XML file format. • Schematics and board layout design software only, no practical simulation. • No maintenance fees and the licensing are really flexible. • Easy to use, easy to learn. •
KiCAD	• Also free and open source. • Just schematic, no simulation tool as eagle. • People use as alternative to eagle. • Has good interface, PCB view is also good, everything well labeled.

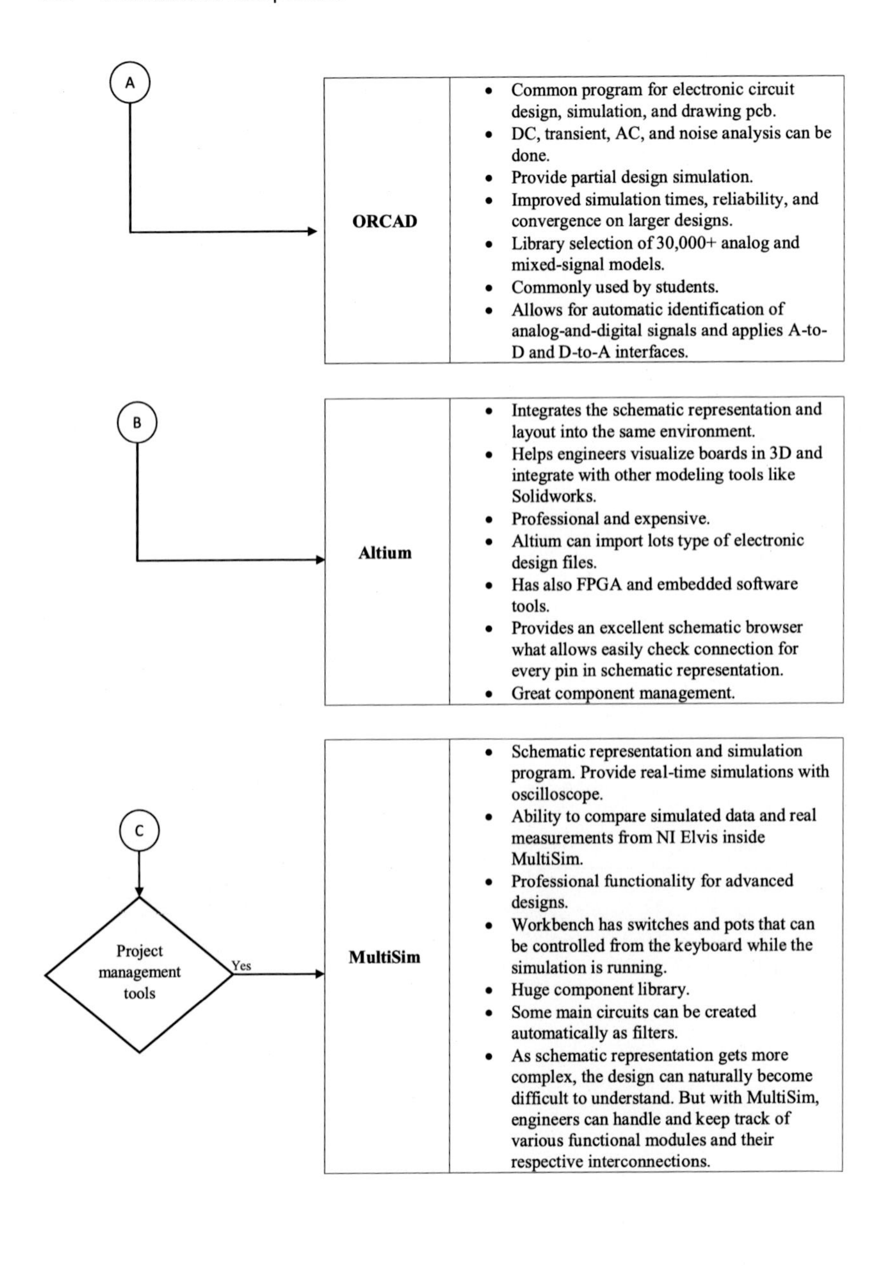
A
ORCAD
Common program for electronic circuit design, simulation, and drawing pcb.
DC, transient, AC, and noise analysis can be done.
Provide partial design simulation.
Improved simulation times, reliability, and convergence on larger designs.
Library selection of 30,000+ analog and mixed-signal models.
Commonly used by students.
Allows for automatic identification of analog-and-digital signals and applies A-to-D and D-to-A interfaces.
B
Altium
Integrates the schematic representation and layout into the same environment.
Helps engineers visualize boards in 3D and integrate with other modeling tools like Solidworks.
Professional and expensive.
Altium can import lots type of electronic design files.
Has also FPGA and embedded software tools.
Provides an excellent schematic browser what allows easily check connection for every pin in schematic representation.
Great component management.
C
Project management tools
Yes
MultiSim
Schematic representation and simulation program. Provide real-time simulations with oscilloscope.
Ability to compare simulated data and real measurements from NI Elvis inside MultiSim.
Professional functionality for advanced designs.
Workbench has switches and pots that can be controlled from the keyboard while the simulation is running.
Huge component library.
Some main circuits can be created automatically as filters.
As schematic representation gets more complex, the design can naturally become difficult to understand. But with MultiSim, engineers can handle and keep track of various functional modules and their respective interconnections.

Chapter 9

Assembly Processes

ABSTRACT

The assembly process can be done temporarily with fasteners or permanently by welding or gluing. If the assembled part requires some kind of service, it is better to connect temporarily. During the assembly process, the order should also be taken into account during the design stage. The assembly process also exists in the electronics. In the prototyping level it is done by hand, but in the commercial level it should be done by automation because the commercial product should be given a warranty at least for 2 years. Moreover, for impact and vibration resistance, it is important that the electronic components are assembled well. It is also very important, from the electric signal perspective, that the soldering of the electronic components is uniform, so that the connection points will not show resistance and heat up.

Mechatronic Components. https://doi.org/10.1016/B978-0-12-814126-7.00009-8

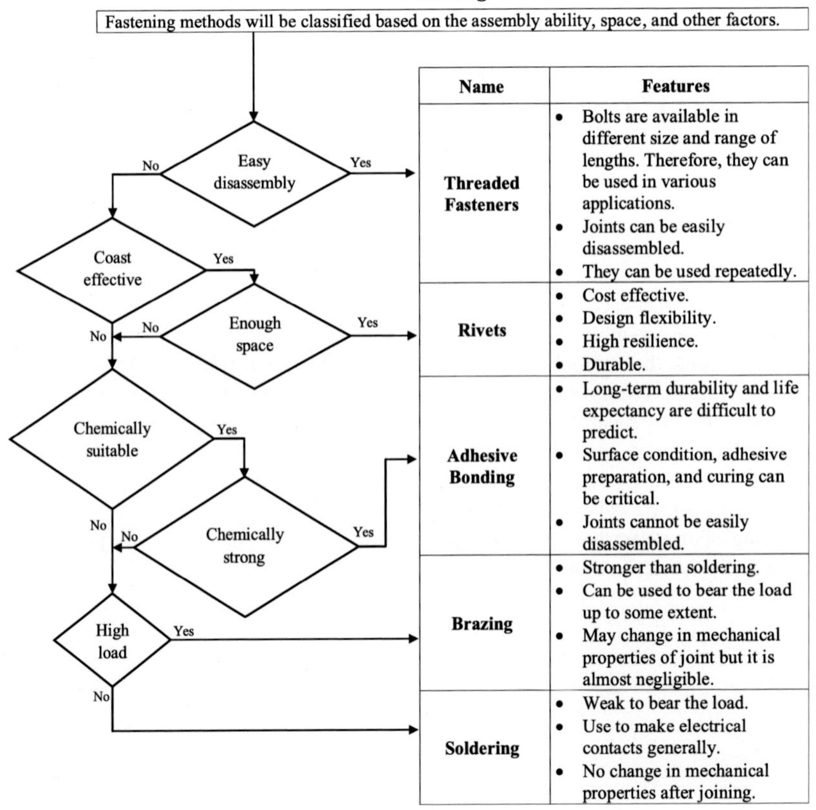

Name	Features
Threaded Fasteners	• Bolts are available in different size and range of lengths. Therefore, they can be used in various applications. • Joints can be easily disassembled. • They can be used repeatedly.
Rivets	• Cost effective. • Design flexibility. • High resilience. • Durable.
Adhesive Bonding	• Long-term durability and life expectancy are difficult to predict. • Surface condition, adhesive preparation, and curing can be critical. • Joints cannot be easily disassembled.
Brazing	• Stronger than soldering. • Can be used to bear the load up to some extent. • May change in mechanical properties of joint but it is almost negligible.
Soldering	• Weak to bear the load. • Use to make electrical contacts generally. • No change in mechanical properties after joining.

Welding

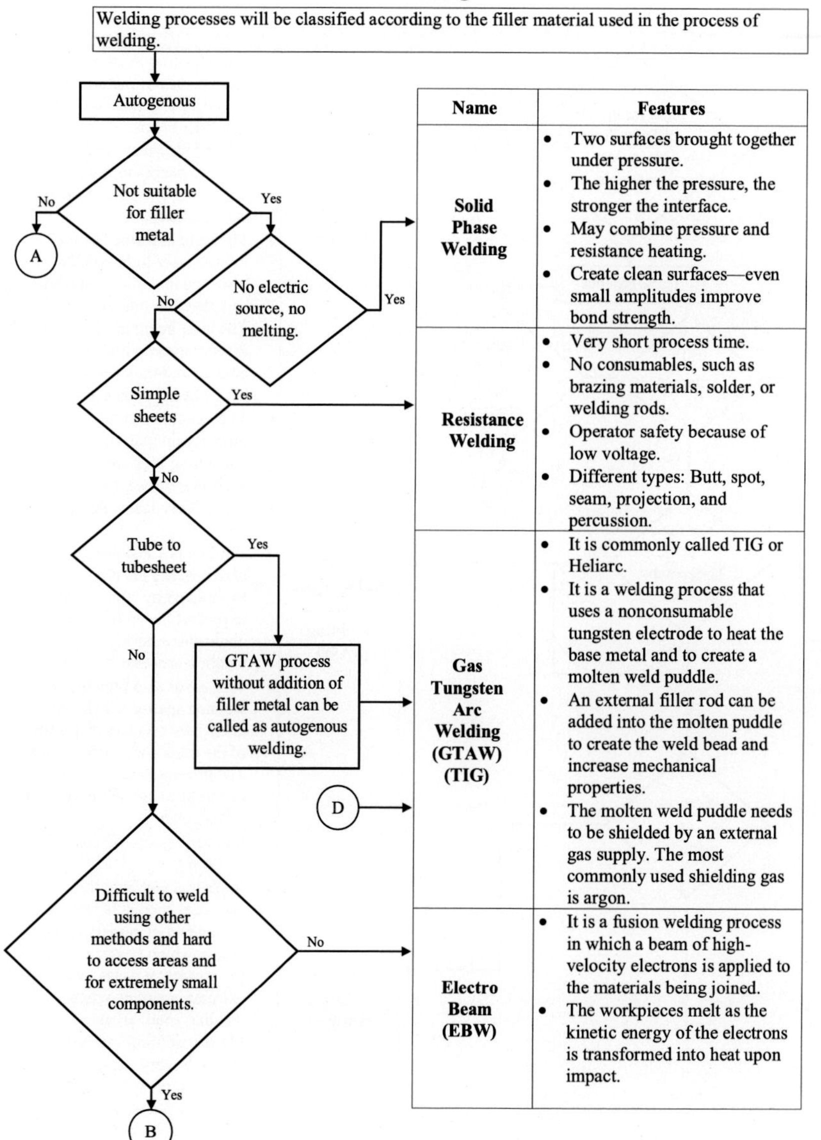

Name	Features
Solid Phase Welding	• Two surfaces brought together under pressure. • The higher the pressure, the stronger the interface. • May combine pressure and resistance heating. • Create clean surfaces—even small amplitudes improve bond strength.
Resistance Welding	• Very short process time. • No consumables, such as brazing materials, solder, or welding rods. • Operator safety because of low voltage. • Different types: Butt, spot, seam, projection, and percussion.
Gas Tungsten Arc Welding (GTAW) (TIG)	• It is commonly called TIG or Heliarc. • It is a welding process that uses a nonconsumable tungsten electrode to heat the base metal and to create a molten weld puddle. • An external filler rod can be added into the molten puddle to create the weld bead and increase mechanical properties. • The molten weld puddle needs to be shielded by an external gas supply. The most commonly used shielding gas is argon.
Electro Beam (EBW)	• It is a fusion welding process in which a beam of high-velocity electrons is applied to the materials being joined. • The workpieces melt as the kinetic energy of the electrons is transformed into heat upon impact.

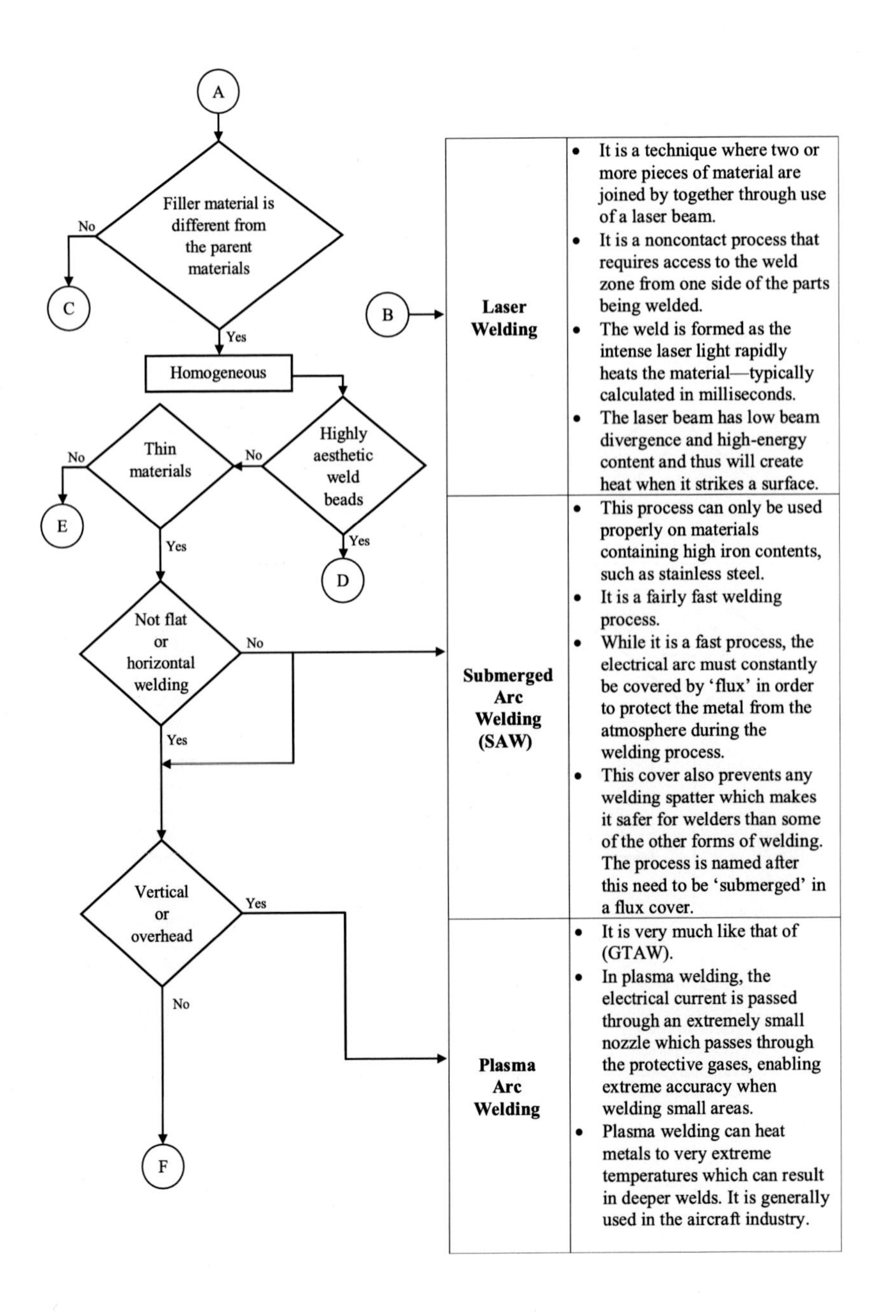

A
Filler material is different from the parent materials
No
C
Yes
Homogeneous
Highly aesthetic weld beads
No
Thin materials
No
E
Yes
Yes
D
B
Not flat or horizontal welding
No
Yes
Vertical or overhead
Yes
No
F
Laser Welding
• It is a technique where two or more pieces of material are joined by together through use of a laser beam.
• It is a noncontact process that requires access to the weld zone from one side of the parts being welded.
• The weld is formed as the intense laser light rapidly heats the material—typically calculated in milliseconds.
• The laser beam has low beam divergence and high-energy content and thus will create heat when it strikes a surface.
Submerged Arc Welding (SAW)
• This process can only be used properly on materials containing high iron contents, such as stainless steel.
• It is a fairly fast welding process.
• While it is a fast process, the electrical arc must constantly be covered by 'flux' in order to protect the metal from the atmosphere during the welding process.
• This cover also prevents any welding spatter which makes it safer for welders than some of the other forms of welding. The process is named after this need to be 'submerged' in a flux cover.
Plasma Arc Welding
• It is very much like that of (GTAW).
• In plasma welding, the electrical current is passed through an extremely small nozzle which passes through the protective gases, enabling extreme accuracy when welding small areas.
• Plasma welding can heat metals to very extreme temperatures which can result in deeper welds. It is generally used in the aircraft industry.

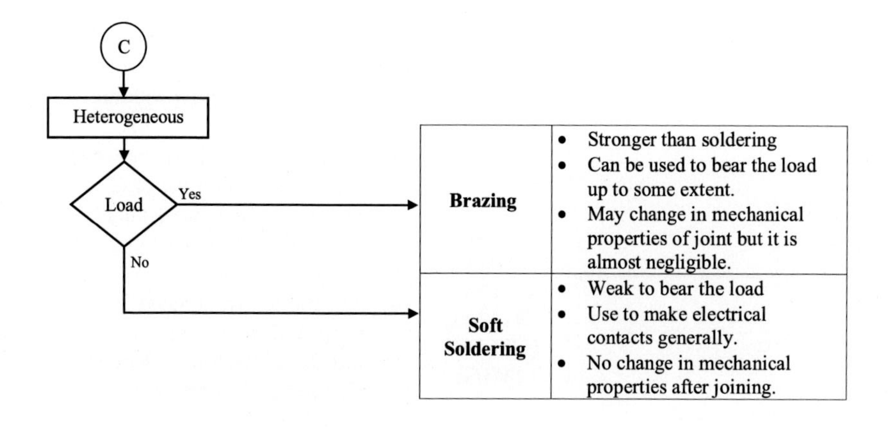
C
Heterogeneous
Load
Yes
No
Brazing
Stronger than soldering
Can be used to bear the load up to some extent.
May change in mechanical properties of joint but it is almost negligible.
Soft Soldering
Weak to bear the load
Use to make electrical contacts generally.
No change in mechanical properties after joining.

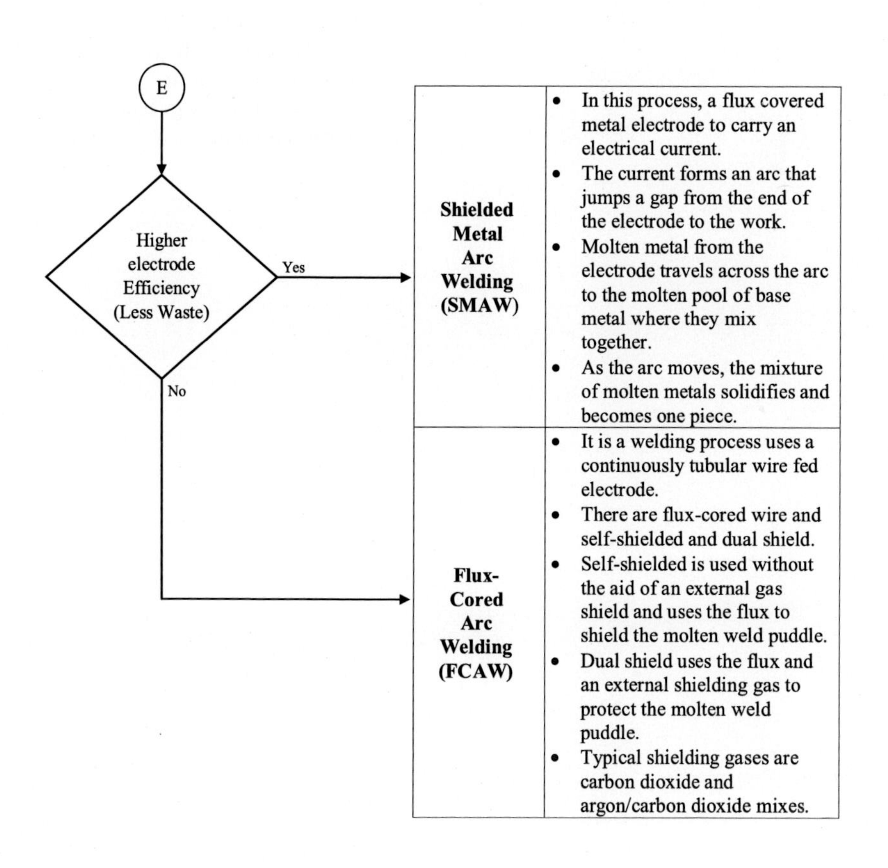
E
Higher electrode Efficiency (Less Waste)
Yes
No
Shielded Metal Arc Welding (SMAW)
In this process, a flux covered metal electrode to carry an electrical current.
The current forms an arc that jumps a gap from the end of the electrode to the work.
Molten metal from the electrode travels across the arc to the molten pool of base metal where they mix together.
As the arc moves, the mixture of molten metals solidifies and becomes one piece.
Flux-Cored Arc Welding (FCAW)
It is a welding process uses a continuously tubular wire fed electrode.
There are flux-cored wire and self-shielded and dual shield.
Self-shielded is used without the aid of an external gas shield and uses the flux to shield the molten weld puddle.
Dual shield uses the flux and an external shielding gas to protect the molten weld puddle.
Typical shielding gases are carbon dioxide and argon/carbon dioxide mixes.

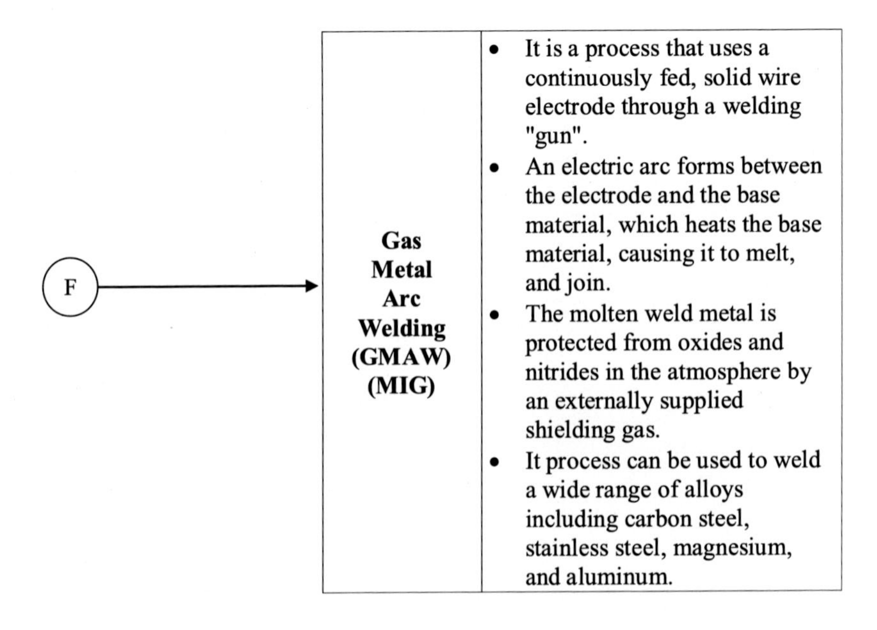
F
Gas
Metal
Arc
Welding
(GMAW)
(MIG)
It is a process that uses a continuously fed, solid wire electrode through a welding "gun".
An electric arc forms between the electrode and the base material, which heats the base material, causing it to melt, and join.
The molten weld metal is protected from oxides and nitrides in the atmosphere by an externally supplied shielding gas.
It process can be used to weld a wide range of alloys including carbon steel, stainless steel, magnesium, and aluminum.

Brazing Process Selection

Brazing is a metal-joining process in which two or more metal items are joined together by melting and flowing a filler metal into the joint, the filler metal having a lower melting point than the adjoining metal. It is classified here based on production volume.

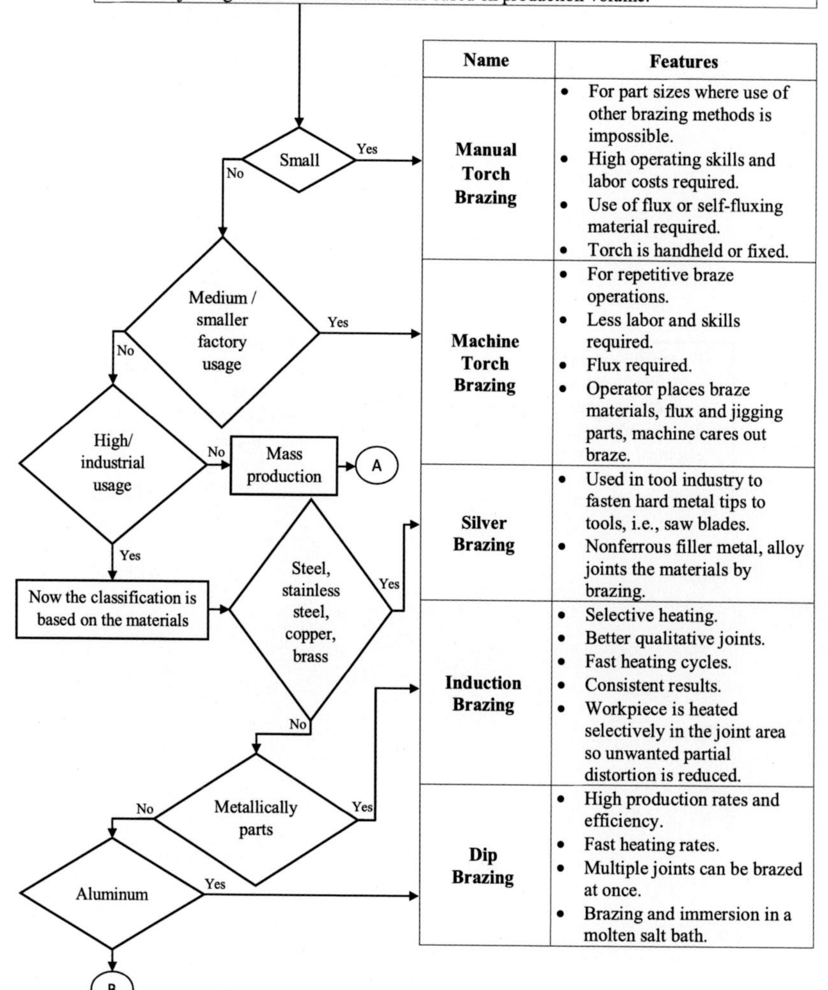

Name	Features
Manual Torch Brazing	• For part sizes where use of other brazing methods is impossible. • High operating skills and labor costs required. • Use of flux or self-fluxing material required. • Torch is handheld or fixed.
Machine Torch Brazing	• For repetitive braze operations. • Less labor and skills required. • Flux required. • Operator places braze materials, flux and jigging parts, machine cares out braze.
Silver Brazing	• Used in tool industry to fasten hard metal tips to tools, i.e., saw blades. • Nonferrous filler metal, alloy joints the materials by brazing.
Induction Brazing	• Selective heating. • Better qualitative joints. • Fast heating cycles. • Consistent results. • Workpiece is heated selectively in the joint area so unwanted partial distortion is reduced.
Dip Brazing	• High production rates and efficiency. • Fast heating rates. • Multiple joints can be brazed at once. • Brazing and immersion in a molten salt bath.

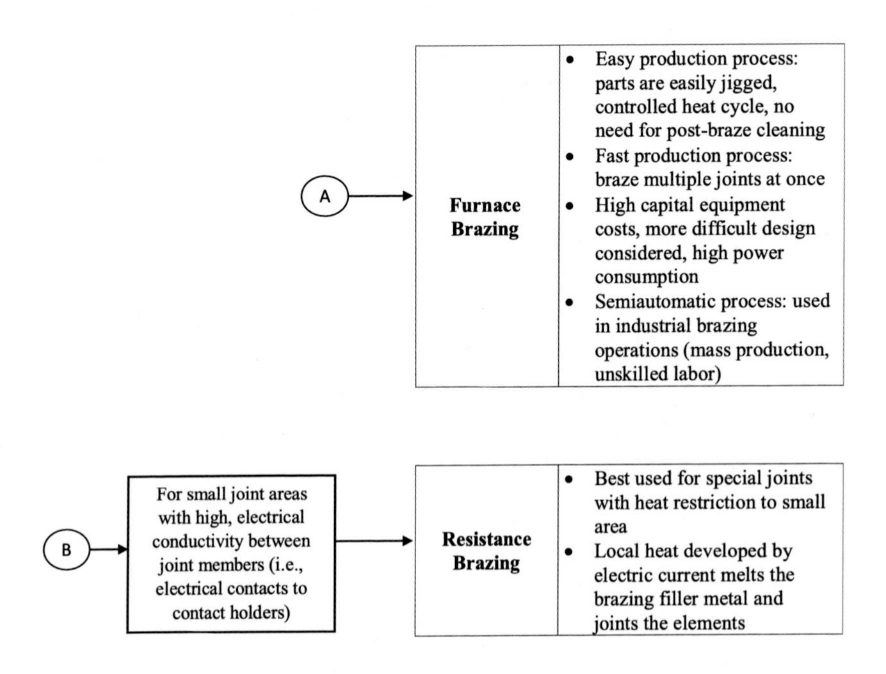
A
Furnace Brazing
Easy production process: parts are easily jigged, controlled heat cycle, no need for post-braze cleaning
Fast production process: braze multiple joints at once
High capital equipment costs, more difficult design considered, high power consumption
Semiautomatic process: used in industrial brazing operations (mass production, unskilled labor)
B
For small joint areas with high, electrical conductivity between joint members (i.e., electrical contacts to contact holders)
Resistance Brazing
Best used for special joints with heat restriction to small area
Local heat developed by electric current melts the brazing filler metal and joints the elements

Soldering

Soldering is a process in which two or more items are joined together by melting and putting a filler metal into the joint, the filler metal having a lower melting point than the adjoining metal.

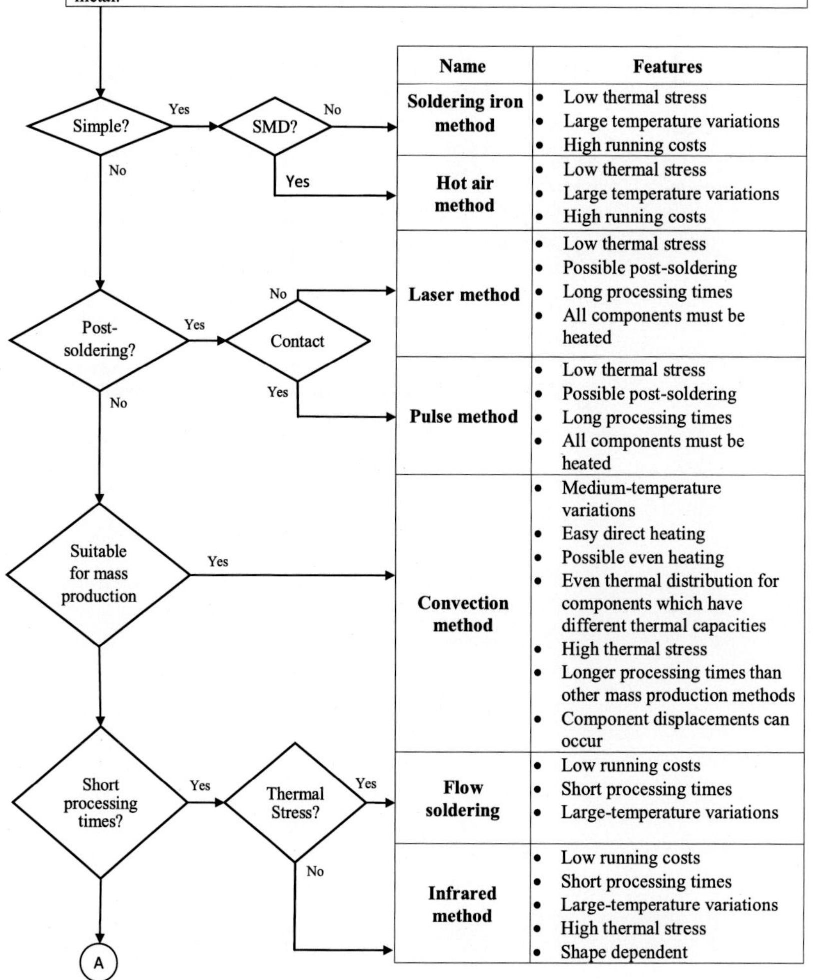

Name	Features
Soldering iron method	• Low thermal stress • Large temperature variations • High running costs
Hot air method	• Low thermal stress • Large temperature variations • High running costs
Laser method	• Low thermal stress • Possible post-soldering • Long processing times • All components must be heated
Pulse method	• Low thermal stress • Possible post-soldering • Long processing times • All components must be heated
Convection method	• Medium-temperature variations • Easy direct heating • Possible even heating • Even thermal distribution for components which have different thermal capacities • High thermal stress • Longer processing times than other mass production methods • Component displacements can occur
Flow soldering	• Low running costs • Short processing times • Large-temperature variations
Infrared method	• Low running costs • Short processing times • Large-temperature variations • High thermal stress • Shape dependent

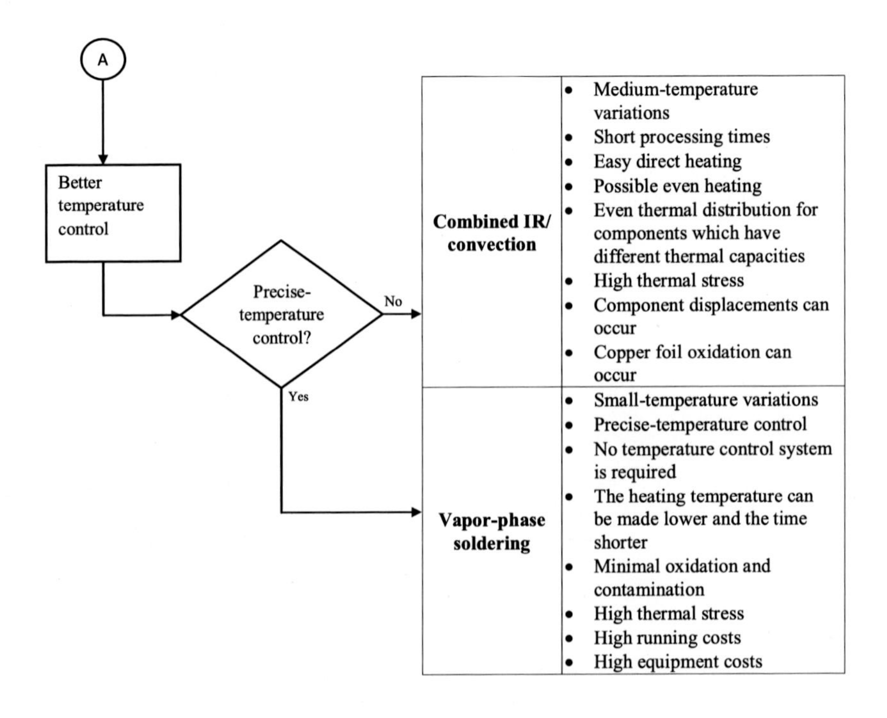
A
Better temperature control
Precise-temperature control?
No
Yes
Combined IR/ convection
Medium-temperature variations
Short processing times
Easy direct heating
Possible even heating
Even thermal distribution for components which have different thermal capacities
High thermal stress
Component displacements can occur
Copper foil oxidation can occur
Vapor-phase soldering
Small-temperature variations
Precise-temperature control
No temperature control system is required
The heating temperature can be made lower and the time shorter
Minimal oxidation and contamination
High thermal stress
High running costs
High equipment costs

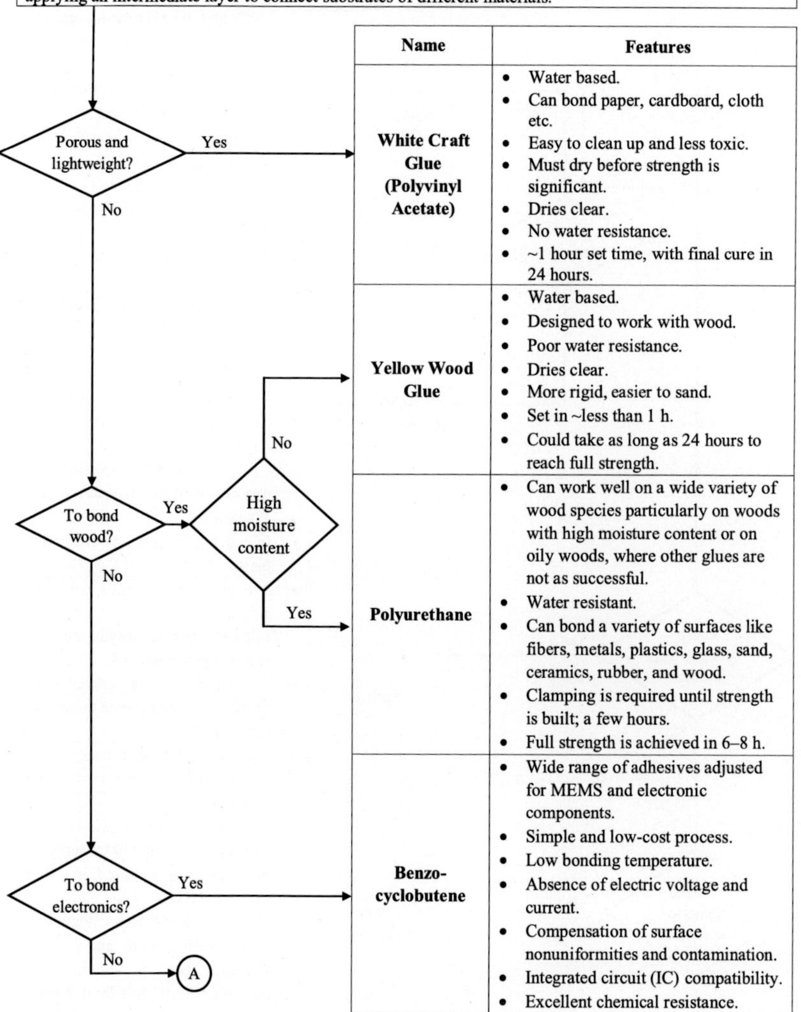
Adhesive Bonding
Adhesive bonding (also referred to as gluing or glue bonding) describes a wafer bonding technique with applying an intermediate layer to connect substrates of different materials.
Porous and lightweight?
Yes
No
To bond wood?
Yes
No
High moisture content
No
Yes
To bond electronics?
Yes
No
A
Name
Features
White Craft Glue (Polyvinyl Acetate)
Water based.
Can bond paper, cardboard, cloth etc.
Easy to clean up and less toxic.
Must dry before strength is significant.
Dries clear.
No water resistance.
~1 hour set time, with final cure in 24 hours.
Yellow Wood Glue
Water based.
Designed to work with wood.
Poor water resistance.
Dries clear.
More rigid, easier to sand.
Set in ~less than 1 h.
Could take as long as 24 hours to reach full strength.
Polyurethane
Can work well on a wide variety of wood species particularly on woods with high moisture content or on oily woods, where other glues are not as successful.
Water resistant.
Can bond a variety of surfaces like fibers, metals, plastics, glass, sand, ceramics, rubber, and wood.
Clamping is required until strength is built; a few hours.
Full strength is achieved in 6–8 h.
Benzo-cyclobutene
Wide range of adhesives adjusted for MEMS and electronic components.
Simple and low-cost process.
Low bonding temperature.
Absence of electric voltage and current.
Compensation of surface nonuniformities and contamination.
Integrated circuit (IC) compatibility.
Excellent chemical resistance.

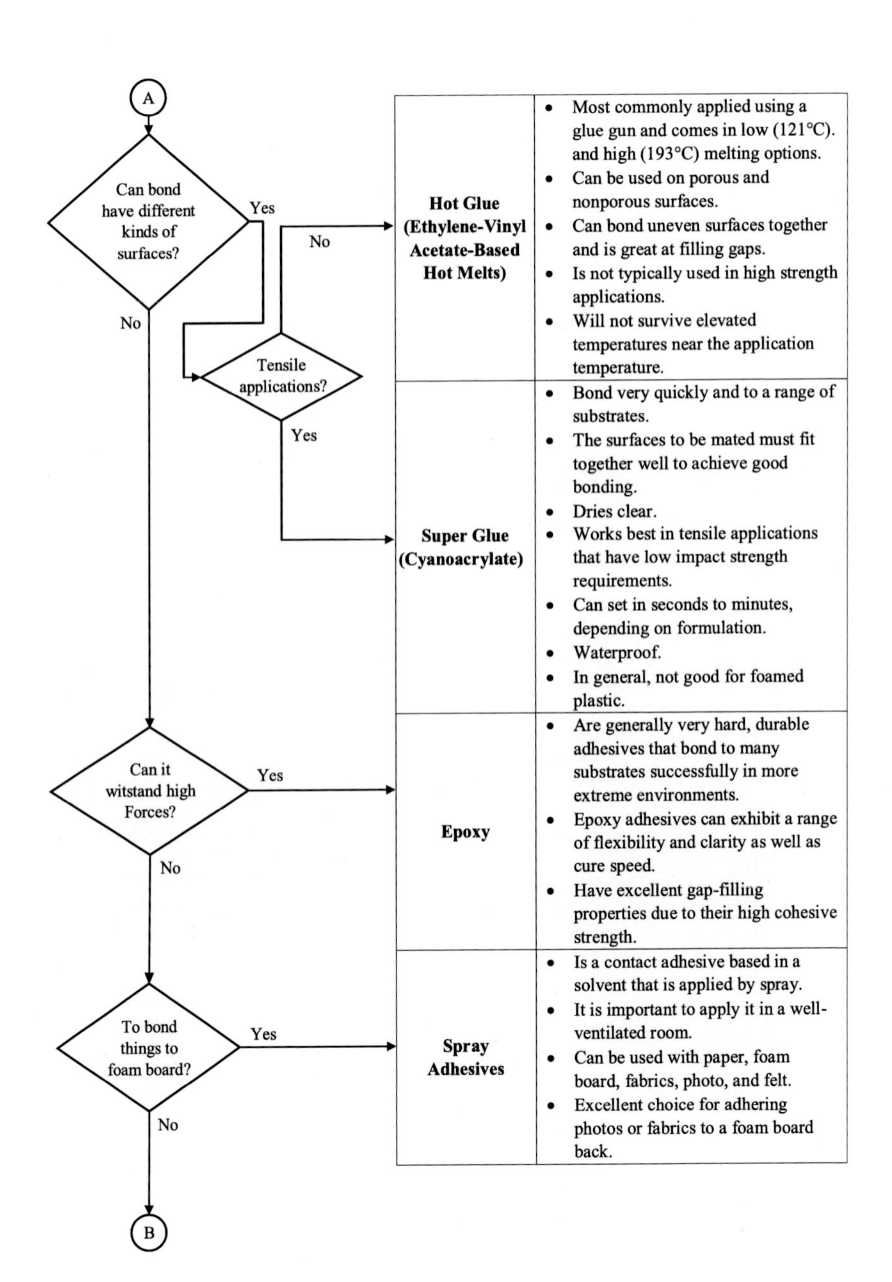

Hot Glue (Ethylene-Vinyl Acetate-Based Hot Melts)	• Most commonly applied using a glue gun and comes in low (121°C). and high (193°C) melting options. • Can be used on porous and nonporous surfaces. • Can bond uneven surfaces together and is great at filling gaps. • Is not typically used in high strength applications. • Will not survive elevated temperatures near the application temperature.
Super Glue (Cyanoacrylate)	• Bond very quickly and to a range of substrates. • The surfaces to be mated must fit together well to achieve good bonding. • Dries clear. • Works best in tensile applications that have low impact strength requirements. • Can set in seconds to minutes, depending on formulation. • Waterproof. • In general, not good for foamed plastic.
Epoxy	• Are generally very hard, durable adhesives that bond to many substrates successfully in more extreme environments. • Epoxy adhesives can exhibit a range of flexibility and clarity as well as cure speed. • Have excellent gap-filling properties due to their high cohesive strength.
Spray Adhesives	• Is a contact adhesive based in a solvent that is applied by spray. • It is important to apply it in a well-ventilated room. • Can be used with paper, foam board, fabrics, photo, and felt. • Excellent choice for adhering photos or fabrics to a foam board back.

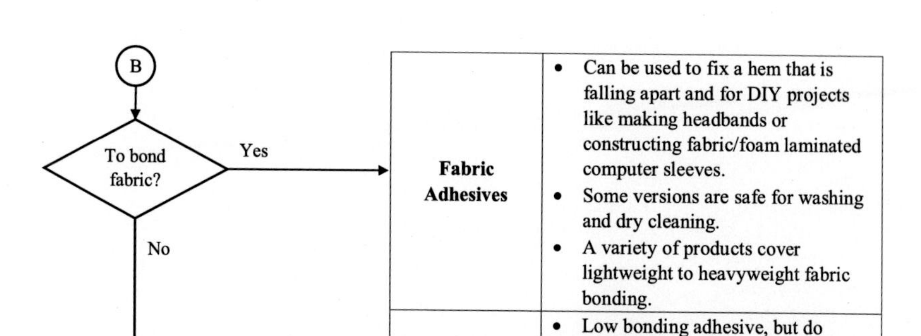
B
To bond fabric?
Yes
No
Fabric Adhesives
Can be used to fix a hem that is falling apart and for DIY projects like making headbands or constructing fabric/foam laminated computer sleeves.
Some versions are safe for washing and dry cleaning.
A variety of products cover lightweight to heavyweight fabric bonding.
Glue Sticks
Low bonding adhesive, but do provide a permanent bond on various types of paper to include cardboard, foam board, and poster board.
Dries clear.

Chapter 10

Electronic Components

ABSTRACT

Electronic components are just like machine elements in mechanical design; they are the basic parts of the electronic circuits. Resistors, capacitors, or cables are general parts that are used in almost any electronic circuit. More advanced parts, like motor drivers, are also available on the market as standard units. The necessity of the electronics in mechatronics projects rises from the fact that the machines cannot be controlled mechanically. The electronic hardware both powers the system and allows control algorithms to be implemented.

Mechatronic Components. https://doi.org/10.1016/B978-0-12-814126-7.00010-4

Cable Selection

Cable selection will be done according to the conductive characteristics of a cable.

Step 1	
To Do	**Notes**
Data gathering	The first step is to collate the relevant information that is required to perform the sizing calculation. Typically, you will need to obtain the following data: • Load Details: o Operating voltage and current o Maximum voltage and current • Cable construction. • Installation conditions.

Step 2	
To Do	**Notes**
Cable selection based on current rating	The component parts that make up the cable must be capable of withstanding the temperature rise and heat emanating from the cable. The current carrying capacity of a cable is the maximum current that can flow continuously through a cable without damaging the cable's insulation and other components.

Step 3				
To Do	**Notes**	**Input**	**Formula**	**Output**
Calculating the voltage drop	The voltage drop will depend on two things: • Current flow through the cable—the higher the current flow, the higher the voltage drop. • Impedance of the conductor—the larger the impedance, the higher the voltage drop.	I, Current R, Resistance of the cable	$V_{Drop} = I.R$	V_{Drop}, Voltage drop
		I, Current Z, Impedance	$E = I.Z$	E, Voltage drop in an alternating current circuit

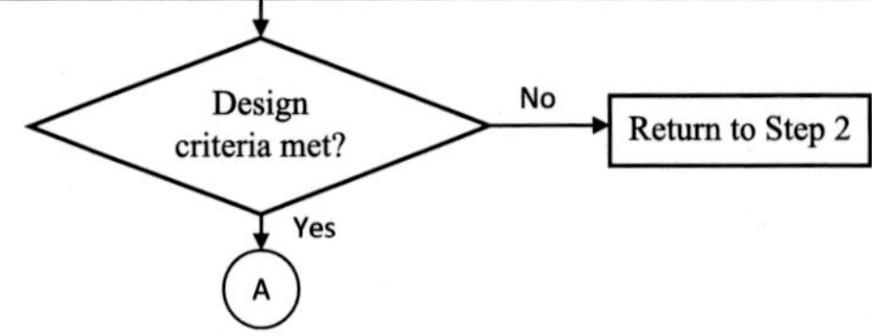

A

Step 4				
To Do	**Notes**	**Input**	**Formula**	**Output**
Calculating the short circuit temperature rise	During a short circuit, a high amount of current can flow through a cable for a short time. This surge in current flow causes a temperature rise within the cable. High temperatures can trigger unwanted reactions in the cable insulation, sheath materials, and other components, which can prematurely degrade the condition of the cable. As the cross-sectional area of the cable increases, it can dissipate higher fault currents for a given temperature rise. Therefore, cables should be sized to withstand the largest short circuit that it is expected to see.	i, Current (A) ρ_d, Density of the cable ($g.mm^{-3}$) ρ_r, Resistivity of the body ($\Omega.mm$) t, Duration of the current flow A, Cross-sectional area of the cable (mm^2) c_p, Heat capacity of the cable	$\Delta T = \frac{i^2 t \rho_r}{A^2 c_p \rho_d}$	ΔT, Adiabatic short circuit temperature rise

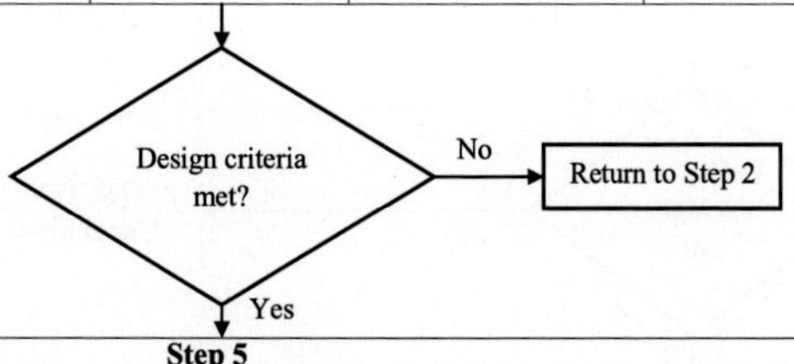

Step 5	
To Do	**Notes**
Earth fault loop impedance	Sometimes it is desirable (or necessary) to consider the earth fault loop impedance of a circuit in the sizing of a cable. Suppose a bolted earth fault occurs between an active conductor and earth. During such an earth fault, it is desirable that the upstream protective device acts to interrupt the fault within a maximum disconnection time to protect against any inadvertent contact to exposed live parts.

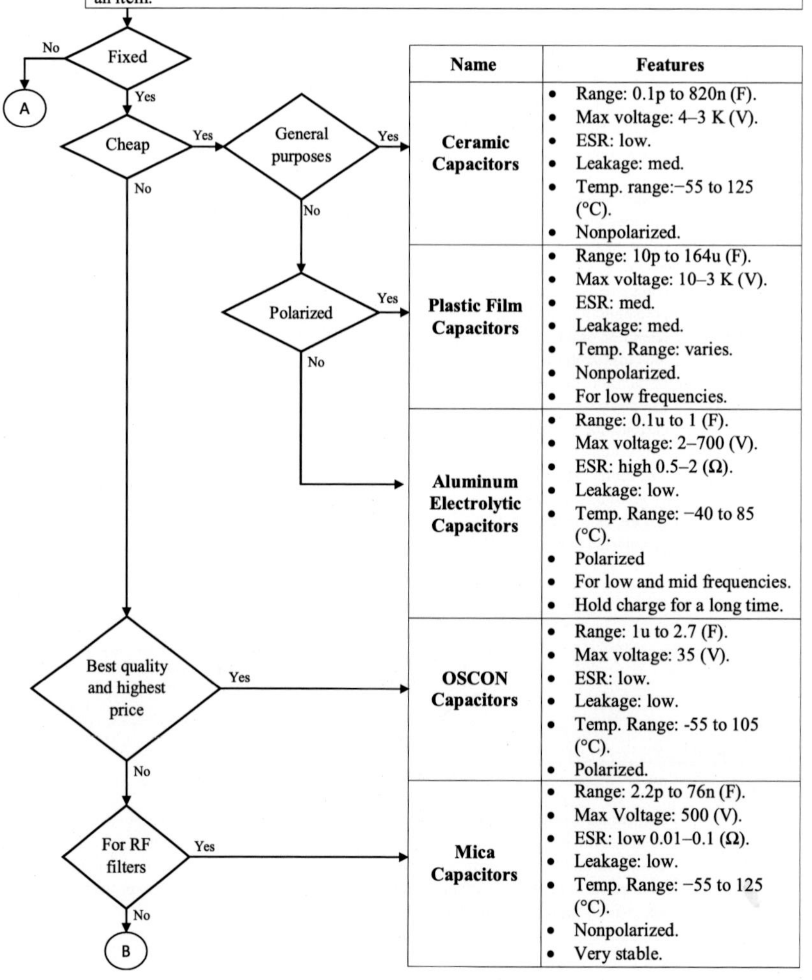

Name	Features
Ceramic Capacitors	• Range: 0.1p to 820n (F). • Max voltage: 4–3 K (V). • ESR: low. • Leakage: med. • Temp. range:−55 to 125 (°C). • Nonpolarized.
Plastic Film Capacitors	• Range: 10p to 164u (F). • Max voltage: 10–3 K (V). • ESR: med. • Leakage: med. • Temp. Range: varies. • Nonpolarized. • For low frequencies.
Aluminum Electrolytic Capacitors	• Range: 0.1u to 1 (F). • Max voltage: 2–700 (V). • ESR: high 0.5–2 (Ω). • Leakage: low. • Temp. Range: −40 to 85 (°C). • Polarized • For low and mid frequencies. • Hold charge for a long time.
OSCON Capacitors	• Range: 1u to 2.7 (F). • Max voltage: 35 (V). • ESR: low. • Leakage: low. • Temp. Range: -55 to 105 (°C). • Polarized.
Mica Capacitors	• Range: 2.2p to 76n (F). • Max Voltage: 500 (V). • ESR: low 0.01–0.1 (Ω). • Leakage: low. • Temp. Range: −55 to 125 (°C). • Nonpolarized. • Very stable.

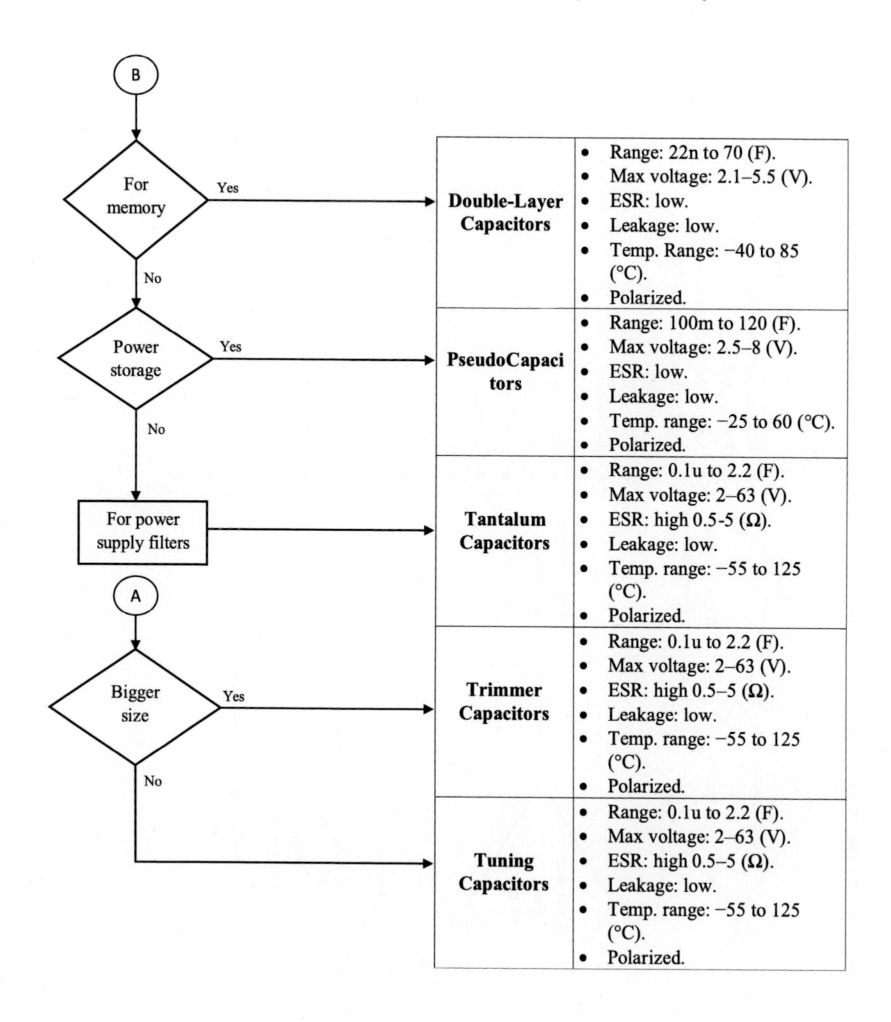
B
For memory
Yes
No
Double-Layer Capacitors
• Range: 22n to 70 (F).
• Max voltage: 2.1–5.5 (V).
• ESR: low.
• Leakage: low.
• Temp. Range: −40 to 85 (°C).
• Polarized.
Power storage
Yes
No
PseudoCapacitors
• Range: 100m to 120 (F).
• Max voltage: 2.5–8 (V).
• ESR: low.
• Leakage: low.
• Temp. range: −25 to 60 (°C).
• Polarized.
For power supply filters
Tantalum Capacitors
• Range: 0.1u to 2.2 (F).
• Max voltage: 2–63 (V).
• ESR: high 0.5-5 (Ω).
• Leakage: low.
• Temp. range: −55 to 125 (°C).
• Polarized.
A
Bigger size
Yes
No
Trimmer Capacitors
• Range: 0.1u to 2.2 (F).
• Max voltage: 2–63 (V).
• ESR: high 0.5–5 (Ω).
• Leakage: low.
• Temp. range: −55 to 125 (°C).
• Polarized.
Tuning Capacitors
• Range: 0.1u to 2.2 (F).
• Max voltage: 2–63 (V).
• ESR: high 0.5–5 (Ω).
• Leakage: low.
• Temp. range: −55 to 125 (°C).
• Polarized.

Resistors

A resistor is a passive two-terminal electrical component that implements electrical resistance as a circuit element.

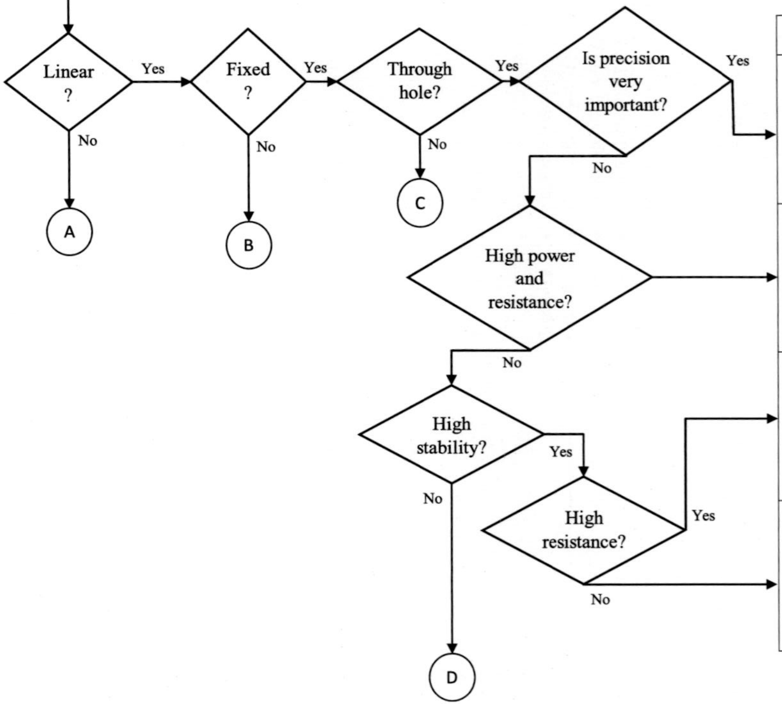

Linear ?
Yes
No
A
Fixed ?
Yes
No
B
Through hole?
Yes
No
C
Is precision very important?
Yes
No
High power and resistance?
No
High stability?
Yes
No
D
High resistance?
Yes
No

Name	Features
Wire Wound Resistors	• Resistance range: 0.1–300 K (Ω). • Rated power: 0.25–1 (W). • Max rated voltage: 25–1000 (V). • Max temperature: (°C). • Tolerance: ±0. 1 to ± 10 (%). • Through hole.
Film Carbon Resistors	• Resistance range: 10–22 M (Ω). • Rated power: 0.25–2 (W). • Max rated voltage:200–1000 (V). • Max temperature: 155 (°C). • Tolerance: ± 2 to ± 10 (%). • Through hole.
Thick Film Resistors	• Resistance range: 1–100 M (Ω). • Rated power: 0.063–0.25 (W). • Max rated voltage: 50–200 (V). • Max temperature: 155 (°C). • Tolerance: ± 1 to ± 5 (%). • Through hole and on-chip.
Thin Film Resistors	• Resistance range: 10–6.19 M (Ω). • Rated power: 50 m to 1 (W). • Max rated voltage: 200 (V). • Max temperature: 155 (°C). • Tolerance: ± 2 to ± 10 (%). • Through hole and on-chip.

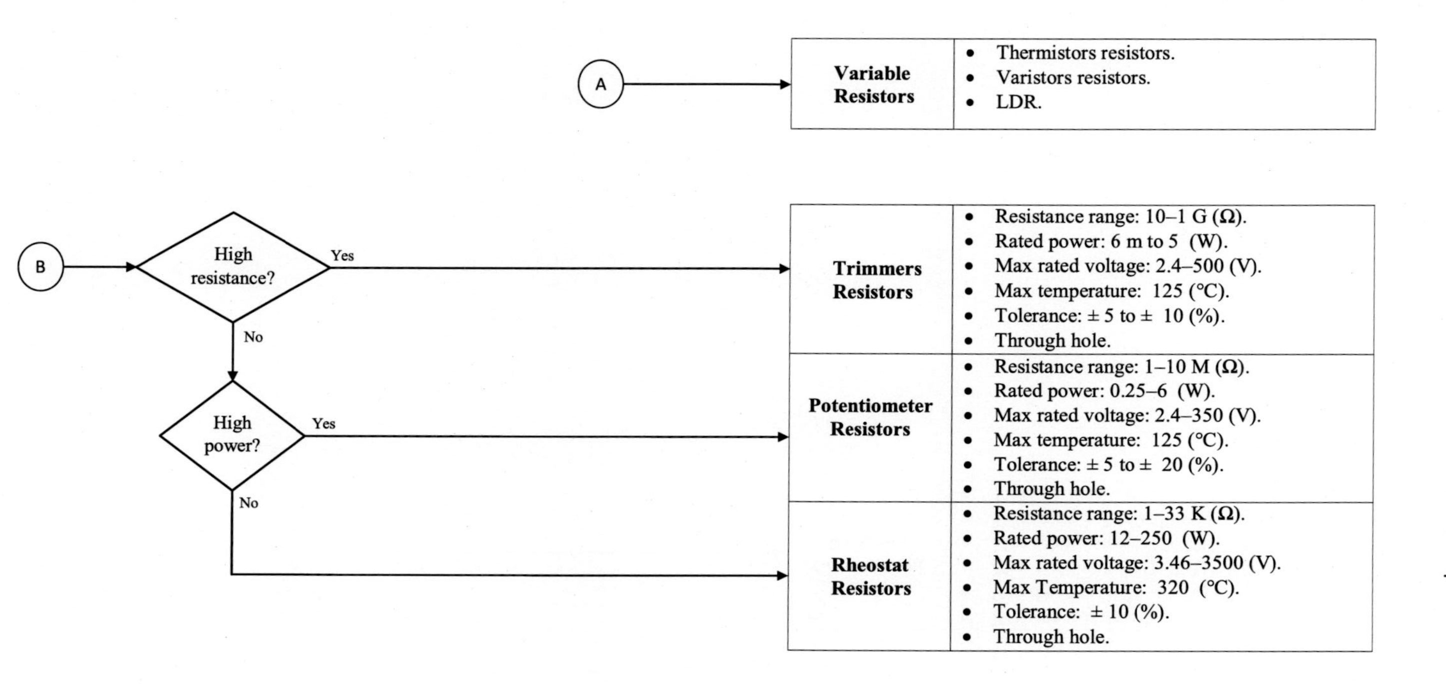
A
Variable Resistors
• Thermistors resistors.
• Varistors resistors.
• LDR.
B
High resistance?
Yes
No
High power?
Yes
No
Trimmers Resistors
• Resistance range: 10–1 G (Ω).
• Rated power: 6 m to 5 (W).
• Max rated voltage: 2.4–500 (V).
• Max temperature: 125 (°C).
• Tolerance: ± 5 to ± 10 (%).
• Through hole.
Potentiometer Resistors
• Resistance range: 1–10 M (Ω).
• Rated power: 0.25–6 (W).
• Max rated voltage: 2.4–350 (V).
• Max temperature: 125 (°C).
• Tolerance: ± 5 to ± 20 (%).
• Through hole.
Rheostat Resistors
• Resistance range: 1–33 K (Ω).
• Rated power: 12–250 (W).
• Max rated voltage: 3.46–3500 (V).
• Max Temperature: 320 (°C).
• Tolerance: ± 10 (%).
• Through hole.

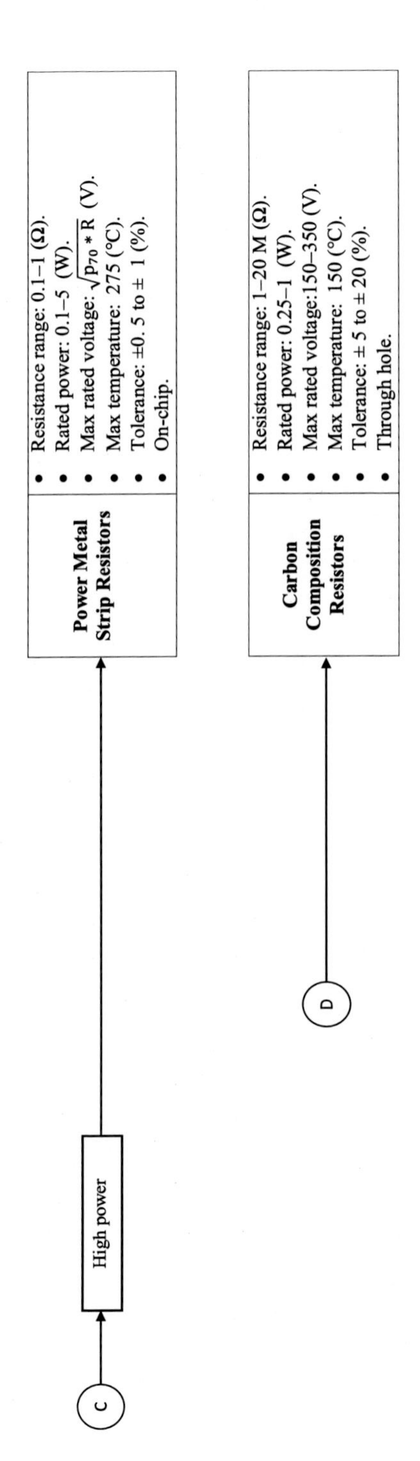
C
High power
Power Metal Strip Resistors
• Resistance range: 0.1–1 (Ω).
• Rated power: 0.1–5 (W).
• Max rated voltage: $\sqrt{p_{70} * R}$ (V).
• Max temperature: 275 (°C).
• Tolerance: ±0. 5 to ± 1 (%).
• On-chip.
D
Carbon Composition Resistors
• Resistance range: 1–20 M (Ω).
• Rated power: 0.25–1 (W).
• Max rated voltage:150–350 (V).
• Max temperature: 150 (°C).
• Tolerance: ± 5 to ± 20 (%).
• Through hole.

Electronic Flow Control

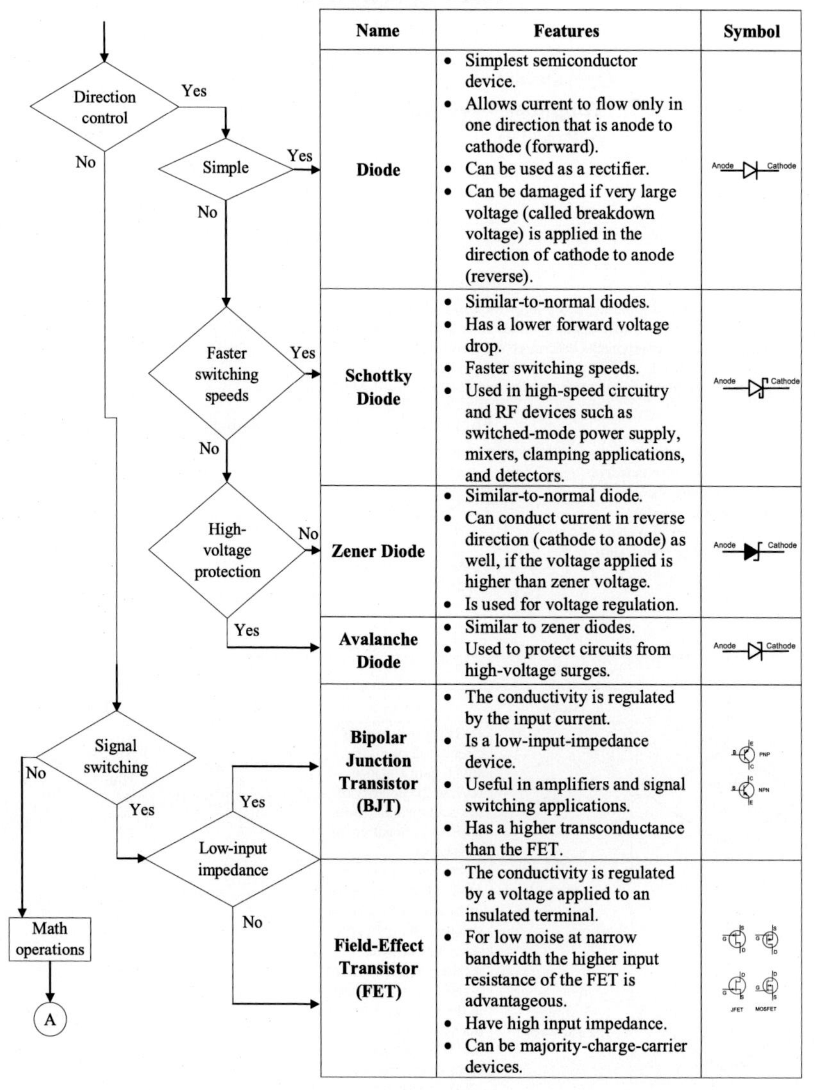

Name	Features	Symbol
Diode	• Simplest semiconductor device. • Allows current to flow only in one direction that is anode to cathode (forward). • Can be used as a rectifier. • Can be damaged if very large voltage (called breakdown voltage) is applied in the direction of cathode to anode (reverse).	Anode Cathode
Schottky Diode	• Similar-to-normal diodes. • Has a lower forward voltage drop. • Faster switching speeds. • Used in high-speed circuitry and RF devices such as switched-mode power supply, mixers, clamping applications, and detectors.	Anode Cathode
Zener Diode	• Similar-to-normal diode. • Can conduct current in reverse direction (cathode to anode) as well, if the voltage applied is higher than zener voltage. • Is used for voltage regulation.	Anode Cathode
Avalanche Diode	• Similar to zener diodes. • Used to protect circuits from high-voltage surges.	Anode Cathode
Bipolar Junction Transistor (BJT)	• The conductivity is regulated by the input current. • Is a low-input-impedance device. • Useful in amplifiers and signal switching applications. • Has a higher transconductance than the FET.	PNP NPN
Field-Effect Transistor (FET)	• The conductivity is regulated by a voltage applied to an insulated terminal. • For low noise at narrow bandwidth the higher input resistance of the FET is advantageous. • Have high input impedance. • Can be majority-charge-carrier devices.	JFET MOSFET

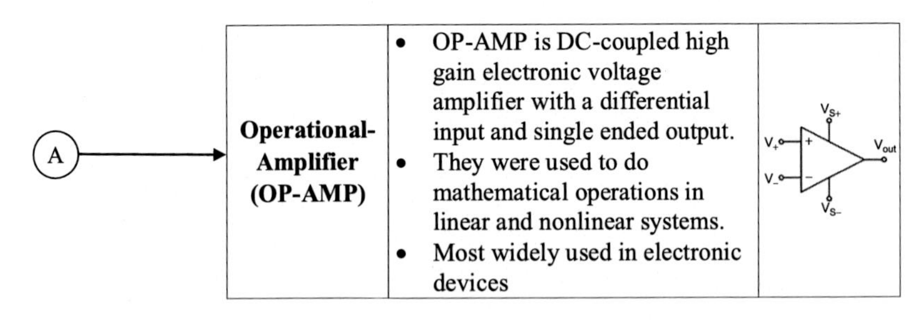

Operational Amplifiers

An Operational Amplifier, or op-amp for short, is fundamentally a voltage amplifying device designed to be used with external feedback components such as resistors and capacitors between its output and input terminals. These feedback components determine the resulting function or "operation" of the amplifier and by virtue of the different feedback configurations whether resistive, capacitive or both, the amplifier can perform a variety of different operations, giving rise to its name of "Operational Amplifier".

Input	Structure	Formula	Output
• The inverting input, marked with a negative or "minus" sign, (−) • The non-inverting Input, marked with a positive or "plus" sign (+).	Non-inverting amplifier: V_{in}, +, −, V_{out}, R_f, R_g	$A_{CL} = 1 + \frac{R_f}{R_g}$	• A_{CL}, Closed-loop gain • V_{out}, Output voltage

Reverse polarity → Yes: Inverting Amplifier; No ↓

Variable voltage → Yes: Voltage Follower; No ↓

Difference → Yes: Difference Amplifier; No → A

Name	Notes	Structure	Formula
Inverting Amplifier	An inverting op amp circuit will boost the signal gain and also reverse the polarity of the output signal, from positive to negative, or vice versa.	R_f, R_{in}, V_{in}, V_{out}	$V_{Out} = -\frac{R_f}{R_{in}} V_{In}$
Voltage Follower	A voltage follower is used to boost the signal of circuits with variable voltages. It applies the same sort of gain increase as the standard amplifier, but variations in the input gain will be tracked and matched by the output signal. These types of circuits are often used as buffers ahead of other systems to prevent damage by sudden changes in voltage.	V_{in}, V_{out}	$V_{Out} = V_{In}$
Difference Amplifier	A difference op-amp outputs a signal that is equal to the difference between its two inputs.	R_f, R_1, R_2, R_g, V_1, V_2, V_{out}	$V_{Out} = \frac{(R_f + R_1)R_g}{(R_g + R_2)R_1} V_2 + \frac{R_f}{R_1} V_1$

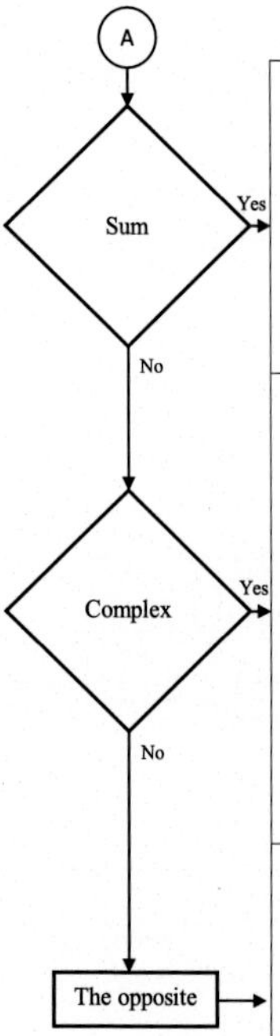

A

Sum — Yes → Summing Amplifiers; No ↓

Complex — Yes → Integrating Amplifier; No ↓

The opposite → Differentiating Amplifier

Type	Description	Circuit	Equation
Summing Amplifiers	A summing amplifier combines different voltages from a number of inputs, and it outputs a gain based on the combined voltages. Either of these circuits can be configured to operate as inverting or non-inverting systems.	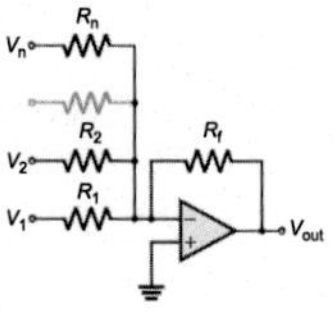	$V_{Out} = -R_f(\frac{V_1}{R_1} + \frac{V_1}{R_1} + \ldots + \frac{V_n}{R_n})$
Integrating Amplifier	The most complex varieties of operational amplifiers are integrators and differentiators. The addition of a capacitor to the circuit means that the integrator reacts to changes in the voltage over time. The magnitude of the output voltage changes based on the amount of time that a voltage spends appearing at the input.	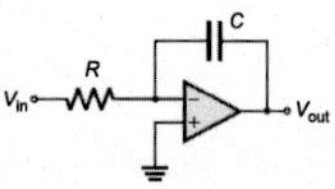	$V_{Out}(t_1) = V_{Out}(t_0) - \frac{1}{RC}\int_{t_0}^{t_1} V_{In}(t)dt$
Differentiating Amplifier	The differentiator is the opposite of the integrator. The voltage produced at the output channel is proportional to the input's rate of change. Greater and faster changes in the input voltage will produce higher output voltages.	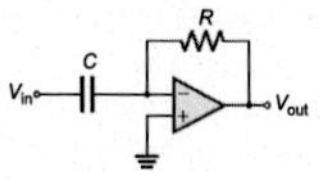	$V_{Out} = -RC\frac{dV_{In}}{dt}$

Logic Operations

Logic operations include any operations that manipulate Boolean values. Boolean values are either true or false.

Name	Description	Truth table
NOT gate	An inverter or NOT gate is a logic gate which implements logical negation. $\hat{A}$	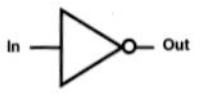 In — NOT — Out In | Out 0 | 1 1 | 0
AND gate	The AND gate is a basic digital logic gate that implements logical conjunction. The function of and effectively finds the minimum between two binary digits. $\boldsymbol{A.B}$	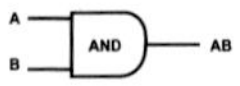 A, B — AND — AB A | B | Out 0 | 0 | 0 0 | 1 | 0 1 | 0 | 0 1 | 1 | 1
OR gate	OR gate is a digital logic gate that implements logical disjunction. The function of OR effectively finds the maximum between two binary digits. $\boldsymbol{A+B}$	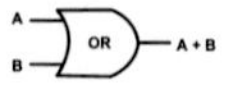 A, B — OR — A+B A | B | Out 0 | 0 | 0 0 | 1 | 1 1 | 0 | 1 1 | 1 | 1
NAND gate	NAND gate (negative-AND) is a logic gate which produces an output which is false only if all its inputs are true; thus its output is complement to that of the AND gate. The NAND gate is significant because any Boolean function can be implemented by using a combination of NAND gates. $\widehat{A.B}$	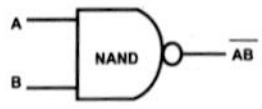 A, B — NAND — $\overline{AB}$ A | B | Out 0 | 0 | 1 0 | 1 | 1 1 | 0 | 1 1 | 1 | 0
NOR gate	NOR is the result of the negation of the OR operator. It can also be seen as an AND gate with all the inputs inverted. NOR gates can be combined to generate any other logical function. $\widehat{A+B}$	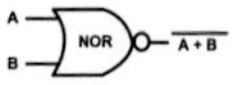 A, B — NOR — $\overline{A+B}$ A | B | Out 0 | 0 | 1 0 | 1 | 0 1 | 0 | 0 1 | 1 | 0
XOR gate	XOR gate is a digital logic gate that implements an exclusive or. XOR represents the inequality function; the output is true if the inputs are not alike, otherwise the output is false. $\hat{A}.B + A.\hat{B}$	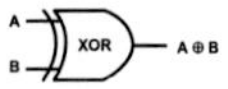 A, B — XOR — $A \oplus B$ A | B | Out 0 | 0 | 0 0 | 1 | 1 1 | 0 | 1 1 | 1 | 0

XNOR gate	The XNOR gate is a digital logic gate whose function is the logical complement of the XOR gate. The two-input version implements logical equality, a HIGH output results if both of the inputs to the gate are the same. $\hat{A}.\hat{B} + A.B$	 A, B, XNOR, $\overline{A \oplus B}$ A | B | Out 0 | 0 | 1 0 | 1 | 0 1 | 0 | 0 1 | 1 | 1
Full Adder	A full adder adds binary numbers and accounts for values carried in as well as out. A one-bit full adder adds three one-bit numbers, often written as *A*, *B*, and C_{in}. A and B are the operands, and C_{in} is a bit carried in from the previous less significant stage. $S = A \oplus B \oplus C_{in}$ $C_{OUT} = (A.B) + (C_{in}.(A \oplus B))$	A, B, Cin, XOR, XOR, Sum, NAND, NAND, NAND, Cout Full Adder

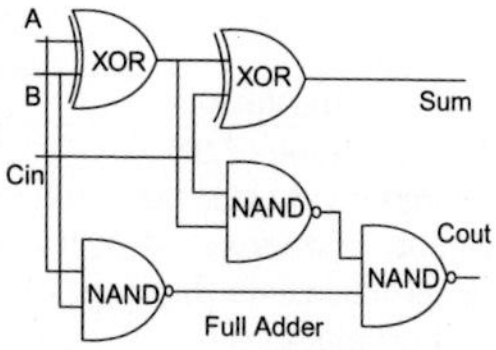

DA/AD Converters

An analog-to-digital converter (ADC, A/D, or A to D) is a device that converts a continuous physical quantity (usually voltage) to a digital number that represents the quantity's amplitude. A digital-to-analog converter (DAC, D/A, D2A, or D-to-A) is a function that converts digital data (usually binary) into an analog signal (current, voltage, or electric charge).

ADC			
Step 1 (Resolution)			
Definition	**Terms**	**Procedure(s)**	
The resolution of the converter indicates the number of discrete values it can produce over the range of analog values.	• E_{FSR}: the full-scale voltage range. • V_{RefHi} and V_{RefLow}: the upper and lower extremes. • Q: Resolution. • M: the ADC's resolution in bits.	1	$E_{FSR} = V_{RefHi} - V_{RefLow}$
		2	$Q = \frac{E_{FSR}}{2^M}$

Step 2 (Quantization Error)		
	Terms	**Procedure(s)**
Quantization error is the noise introduced by quantization in an ideal ADC. It is a rounding error between the analog input voltage to the ADC and the output-digitized value. The noise is nonlinear and signal dependent.	• SQNR: Signal-to-quantization-noise ratio (db). • M: the number of quantization bits.	$SQNR = 6.02 * M$

Step 3 (Dither)
Procedure(s)
This is a very small amount of random noise (white noise), which is added to the input before conversion. That helps for accurate representation of the signal over time.

Step 4 (Accuracy)		
Definition	**Terms**	**Procedure(s)**
An ADC has several sources of errors. Quantization error and (assuming the ADC is intended to be linear) non-linearity are intrinsic to any analog-to-digital conversion	M: the number of quantizatio n bits.	$\frac{1}{2^M}$

A

A

Step 5 (Jitter)		
Definition	**Terms**	**Procedure(s)**
The use of a nonideal sampling clock will result in some uncertainty in when samples are recorded. Provided that the actual sampling time uncertainty due to the clock jitter is Δt. This will result in additional recorded noise that will reduce the effective number of bits.	• Δt: Actual sampling time. • M: the number of quantization bits. • f_0: Signal frequency.	$\Delta t < \frac{1}{2^M \pi f_0}$

Step 6 (Sample Frequency)	
Terms	**Procedure(s)**
• f_s: Sampling frequency. • f: Highest frequency component of the original signal.	$\frac{f_s}{2} > f$

Step 7 (Relative Speed and Precision)
The speed of an ADC varies by type. The Wilkinson ADC is limited by the clock rate, which is processable by current digital circuits. Currently, frequencies up to 300 MHz are possible.

DAC		
Step 1 (Resolution)		
Definition	**Terms**	**Procedure(s)**
The number of possible output levels the DAC is designed to reproduce. This is usually stated as the number of bits it uses, which is the base two logarithm of the number of levels.	M: the number of quantization bits.	2^M

Step 2 (Maximum Sampling Rate)
A measurement of the maximum speed at which the DACs circuitry can operate and still produce the correct output. As stated above, the Nyquist-Shannon sampling theorem defines a relationship between this and the bandwidth of the sampled signal.

Step 3 (Monotonicity)
The ability of a DAC's analog output to move only in the direction that the digital input moves (i.e., if the input increases, the output does not dip before asserting the correct output.) This characteristic is very important for DACs used as a low-frequency signal source or as a digitally programmable trim element.

B

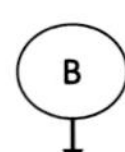

Step 4 [Total Harmonic Distortion and Noise (THD+N)]
A measurement of the distortion and noise introduced to the signal by the DAC. It is expressed as a percentage of the total power of unwanted harmonic distortion and noise that accompany the desired signal. This is a very important DAC characteristic for dynamic and small signal DAC applications.

Step 5 (Dynamic Range)
A measurement of the difference between the largest and smallest signals the DAC can reproduce expressed in decibels. This is usually related to resolution and noise floor. Other measurements, such as phase distortion and jitter, can also be very important for some applications, some of which (e.g., wireless data transmission, composite video) may even rely on accurate production of phase-adjusted signals.

Multiplexers

A multiplexer (or mux) is a device that selects one of several analog or digital input signals and forwards the selected input into a single line.

Step 1	
To Do	**Notes**
Determining whether to use analog or a digital multiplexer	• In analog circuit design, a multiplexer is a special type of analog switch that connects one signal selected from several inputs to a single output. • In digital circuit design, the selector wires are of digital value. In the case of a 2-to-1 multiplexer, a logic value of 0 would connect I_0 to the output while a logic value of 1 would connect I_1 to the output.

Step 2				
To Do	**Notes**	**Input**	**Formula**	**Output**
Determining the number of selector pins	A multiplexer of 2^n inputs has n select lines, which are used to select which input line to send to the output	n, Number of inputs	$x = \log_2 n$	x, Number of select lines

Step 3					
To Do	**Structure**	**Input**	**Formula**	**Table**	**Output**
Finding the truth table	S_0; A — 0; B — 1; Z	A, Input 1 B, Input 2 S, Selector pin	$Z = (A.\bar{S}) + (B.S)$	See table below	Z, Output

S	A	B	Z
0	1	1	1
		0	1
	0	1	0
		0	0
1	1	1	1
		0	0
	0	1	1
		0	0

Bridge Circuits

A bridge circuit is a type of electrical circuit in which two circuit branches are "bridged" by a third branch connected between the first two branches at some intermediate point along them.

Name	Features	Schematics
Wheatstone bridge	A Wheatstone bridge is an electrical circuit used to measure an unknown electrical resistance by balancing two legs of a bridge circuit, one leg of which includes the unknown component.	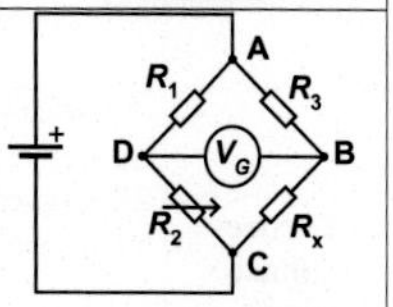
Wien bridge	Wien bridge comprises four resistors and two capacitors. Bridge circuits were a common way of measuring component values by comparing them to known values.	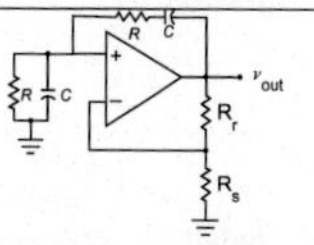
Maxwell bridge	A Maxwell bridge is a modification to a Wheatstone bridge used to measure an unknown inductance in terms of calibrated resistance and capacitance.	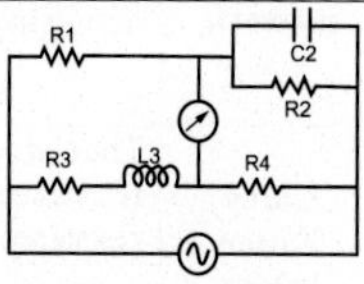
H bridge	A H bridge is an electronic circuit that enables a voltage to be applied across a load in either direction. These circuits are often used in robotics and other applications to allow DC motors to run forwards and backwards.	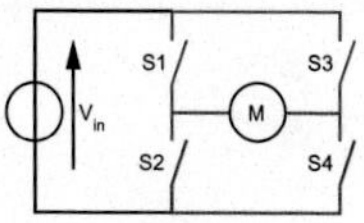
Fontana bridge	A Fontana bridge is a type of bridge circuit that implements a wide-frequency band voltage-to-current converter. The converter is characterized by a combination of positive and negative feedback loops, implicit in this bridge configuration.	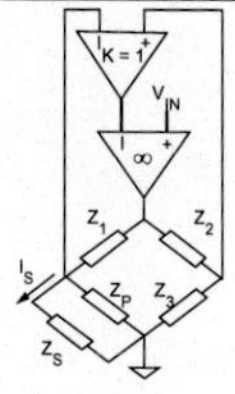
Diode bridge	A diode bridge is an arrangement of four diodes in a bridge circuit configuration that provides the same polarity of output for either polarity of input.	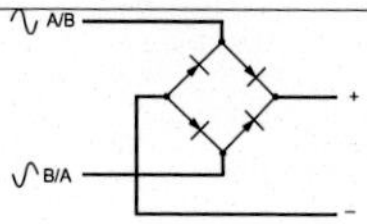

Kelvin bridge	A Kelvin bridge is a measuring instrument used to measure unknown electrical resistors below 1 ohm. It is specifically designed to measure resistors that are constructed as four terminal resistors.	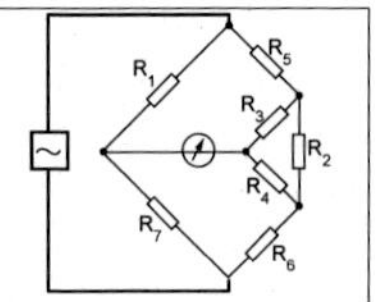
Lattice-phase equalizer	A lattice-phase equalizer or lattice filter is an example of an all-pass filter. That is, the attenuation of the filter is constant at all frequencies but the relative phase between input and output varies with frequency.	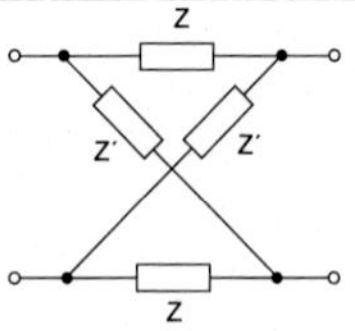
Zobel network	Zobel networks are a type of filter section based on the image-impedance design principle. The distinguishing feature of Zobel networks is that the input impedance is fixed in the design independently of the transfer function.	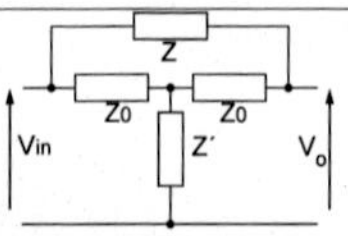
Carey Foster bridge	The Carey Foster bridge is a bridge circuit used to measure low resistances, or to measure small differences between two large resistances.	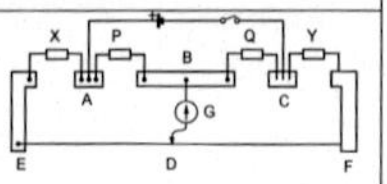

Displays

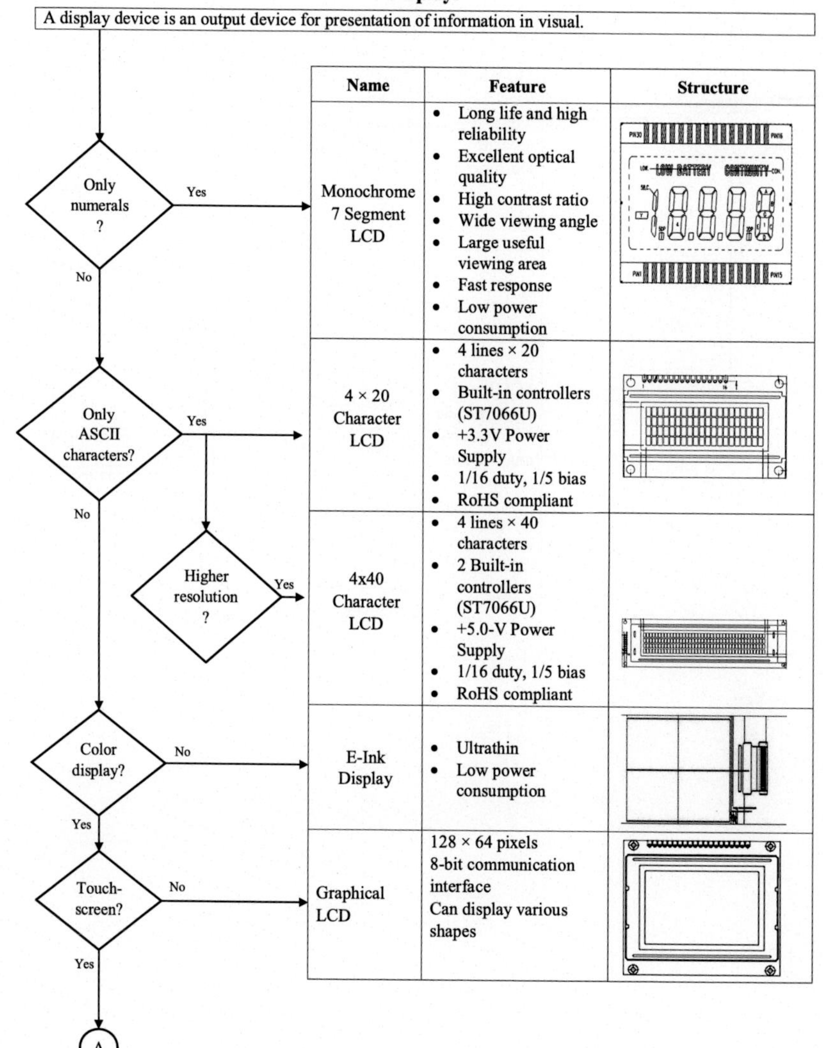

Name	Feature	Structure
Monochrome 7 Segment LCD	• Long life and high reliability • Excellent optical quality • High contrast ratio • Wide viewing angle • Large useful viewing area • Fast response • Low power consumption	
4 × 20 Character LCD	• 4 lines × 20 characters • Built-in controllers (ST7066U) • +3.3V Power Supply • 1/16 duty, 1/5 bias • RoHS compliant	
4x40 Character LCD	• 4 lines × 40 characters • 2 Built-in controllers (ST7066U) • +5.0-V Power Supply • 1/16 duty, 1/5 bias • RoHS compliant	
E-Ink Display	• Ultrathin • Low power consumption	
Graphical LCD	128 × 64 pixels 8-bit communication interface Can display various shapes	

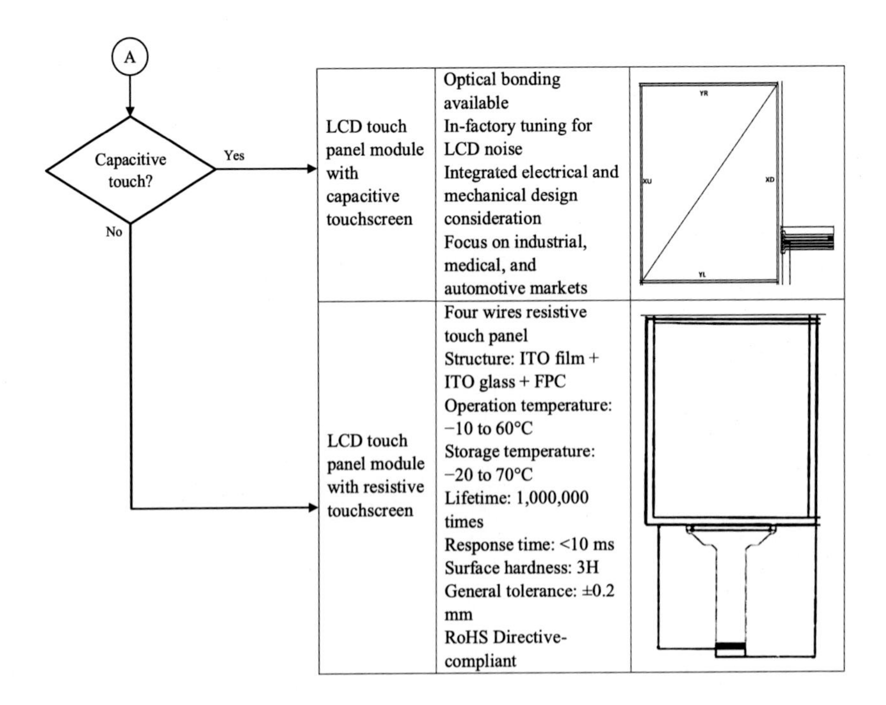
A
Capacitive touch?
Yes
No
LCD touch panel module with capacitive touchscreen
Optical bonding available
In-factory tuning for LCD noise
Integrated electrical and mechanical design consideration
Focus on industrial, medical, and automotive markets
YR
XU
XD
YL
LCD touch panel module with resistive touchscreen
Four wires resistive touch panel
Structure: ITO film + ITO glass + FPC
Operation temperature: −10 to 60°C
Storage temperature: −20 to 70°C
Lifetime: 1,000,000 times
Response time: <10 ms
Surface hardness: 3H
General tolerance: ±0.2 mm
RoHS Directive-compliant

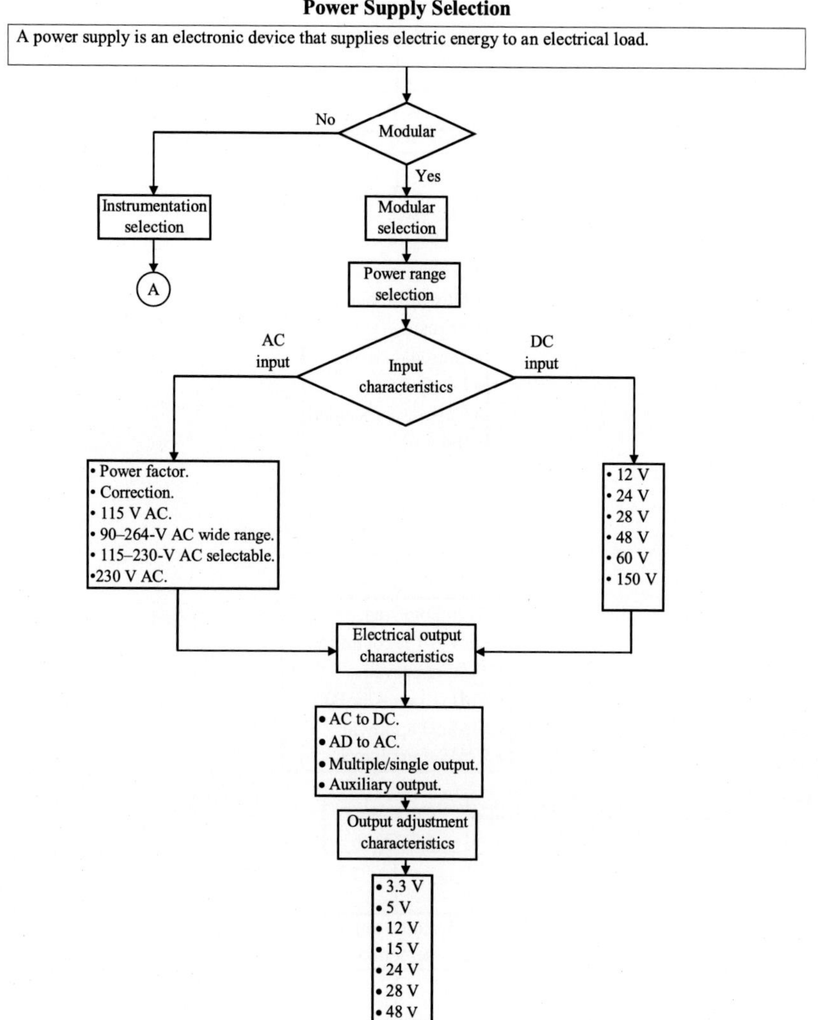
Power Supply Selection
A power supply is an electronic device that supplies electric energy to an electrical load.
Modular
No
Yes
Instrumentation selection
A
Modular selection
Power range selection
AC input
Input characteristics
DC input
• Power factor.
• Correction.
• 115 V AC.
• 90–264-V AC wide range.
• 115–230-V AC selectable.
•230 V AC.
• 12 V
• 24 V
• 28 V
• 48 V
• 60 V
• 150 V
Electrical output characteristics
• AC to DC.
• AD to AC.
• Multiple/single output.
• Auxiliary output.
Output adjustment characteristics
• 3.3 V
• 5 V
• 12 V
• 15 V
• 24 V
• 28 V
• 48 V
B

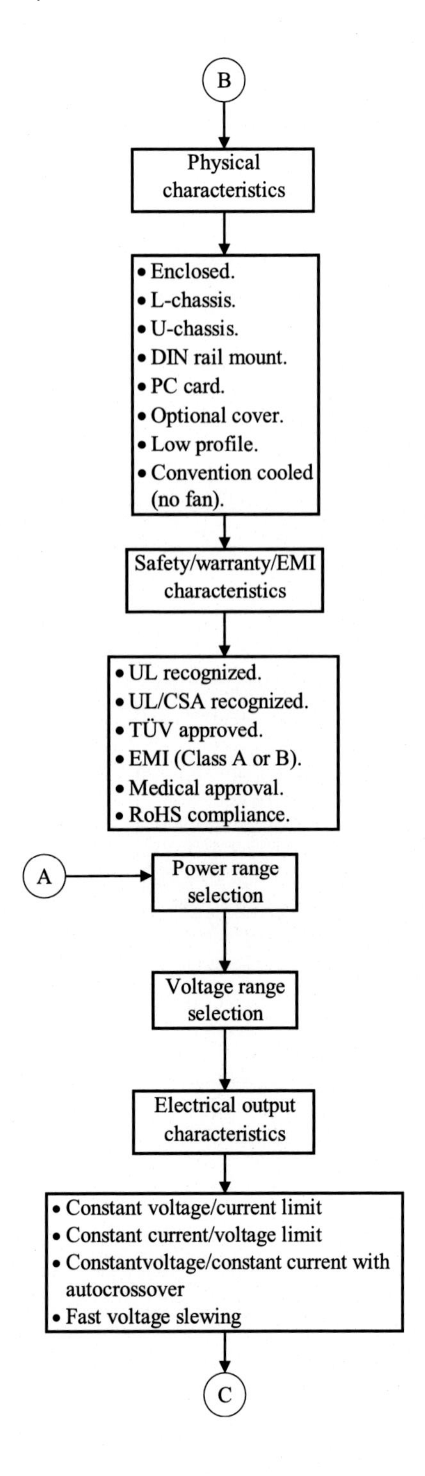
B
Physical characteristics
• Enclosed.
• L-chassis.
• U-chassis.
• DIN rail mount.
• PC card.
• Optional cover.
• Low profile.
• Convention cooled (no fan).
Safety/warranty/EMI characteristics
• UL recognized.
• UL/CSA recognized.
• TÜV approved.
• EMI (Class A or B).
• Medical approval.
• RoHS compliance.
A
Power range selection
Voltage range selection
Electrical output characteristics
• Constant voltage/current limit
• Constant current/voltage limit
• Constantvoltage/constant current with autocrossover
• Fast voltage slewing
C

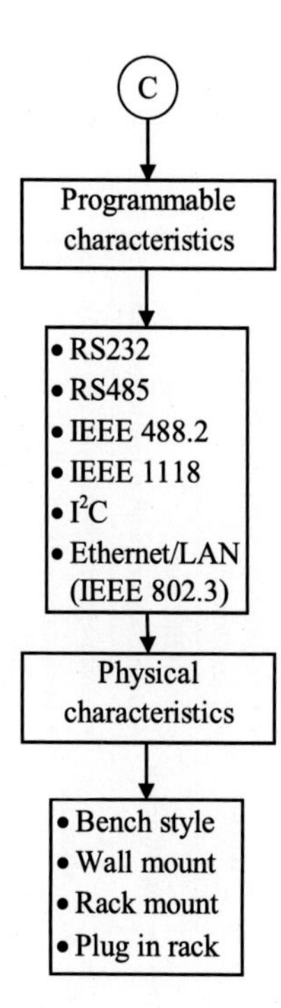
C
Programmable characteristics
• RS232
• RS485
• IEEE 488.2
• IEEE 1118
• I²C
• Ethernet/LAN (IEEE 802.3)
Physical characteristics
• Bench style
• Wall mount
• Rack mount
• Plug in rack

Batteries

An electric battery is a device consisting of two or more electrochemical cells that convert stored chemical energy into electrical energy.

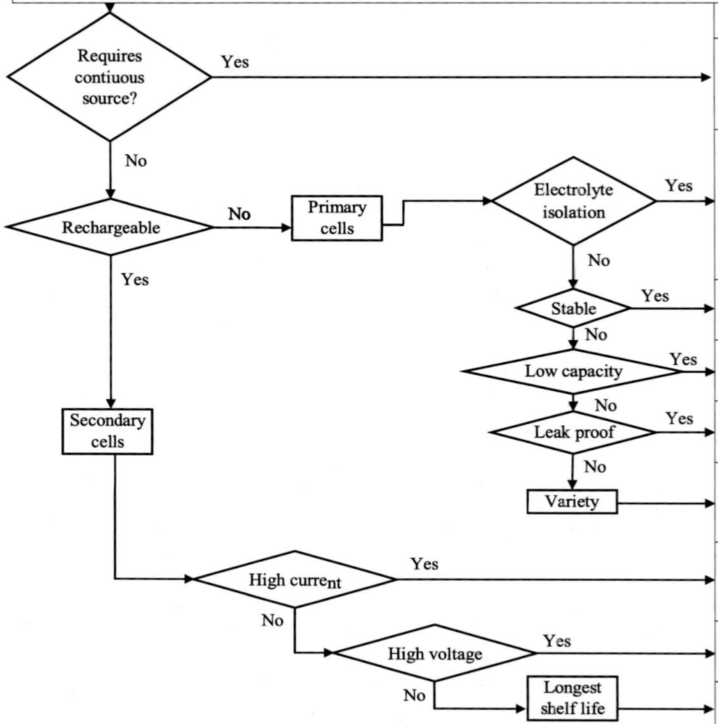

Requires contiuous source?
Yes
No
Rechargeable
No
Yes
Primary cells
Secondary cells
Electrolyte isolation
Yes
No
Stable
Yes
No
Low capacity
Yes
No
Leak proof
Yes
No
Variety
High current
Yes
No
High voltage
Yes
No
Longest shelf life

Name	Features
Fuel Cells	• They are electrochemical devices which convert chemical energy to electrical energy • They can operate continuously as long as input components supplied
Reserve Cells	• Designed for emergency use • Eliminate self-discharge • Usually used in military and marine applications • Electrolyte activated and start energy flow • Thermal, water, and ampoule activated types are most common
Mercury	• Small, steady, and long lifetime • Cause toxicity
Zinc-Carbon	• Short lifetime • Does not work well at low temperatures.
Alkaline	• Most popular dry-cell with low load current • Environmentally friendly
Silver-Oxide	• Small-to-large size • Expensive and better run time than alkaline • Being used in hearing aids
Lead-Acid	• Oldest rechargeable battery • Can be recharged by an alternator • Generally used where high-current need
Nickel-Zinc(NiZn)	• High capacity and voltage • Very short life cycle
Rechargeable Alkaline	• Lowest discharge among secondaries • Poor performance in high-drain devices

Digital Devices

A digital device is an electronic device which uses discrete, numerable data, and processes for all its operations. All digital devices have some sort of combination of mentioned digital devices which are basically building blocks

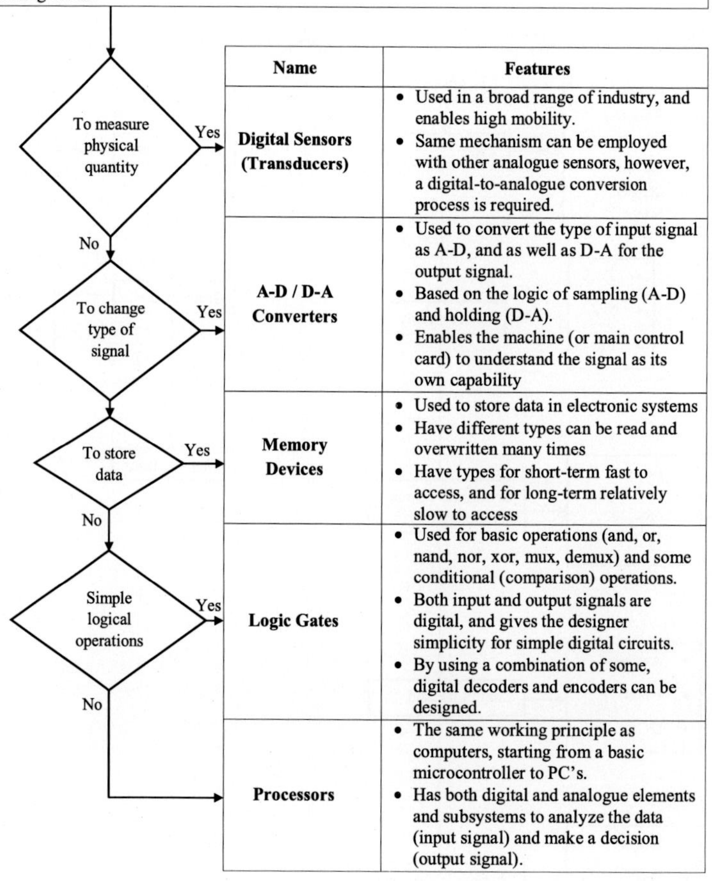

Name	Features
Digital Sensors (Transducers)	• Used in a broad range of industry, and enables high mobility. • Same mechanism can be employed with other analogue sensors, however, a digital-to-analogue conversion process is required.
A-D / D-A Converters	• Used to convert the type of input signal as A-D, and as well as D-A for the output signal. • Based on the logic of sampling (A-D) and holding (D-A). • Enables the machine (or main control card) to understand the signal as its own capability
Memory Devices	• Used to store data in electronic systems • Have different types can be read and overwritten many times • Have types for short-term fast to access, and for long-term relatively slow to access
Logic Gates	• Used for basic operations (and, or, nand, nor, xor, mux, demux) and some conditional (comparison) operations. • Both input and output signals are digital, and gives the designer simplicity for simple digital circuits. • By using a combination of some, digital decoders and encoders can be designed.
Processors	• The same working principle as computers, starting from a basic microcontroller to PC's. • Has both digital and analogue elements and subsystems to analyze the data (input signal) and make a decision (output signal).

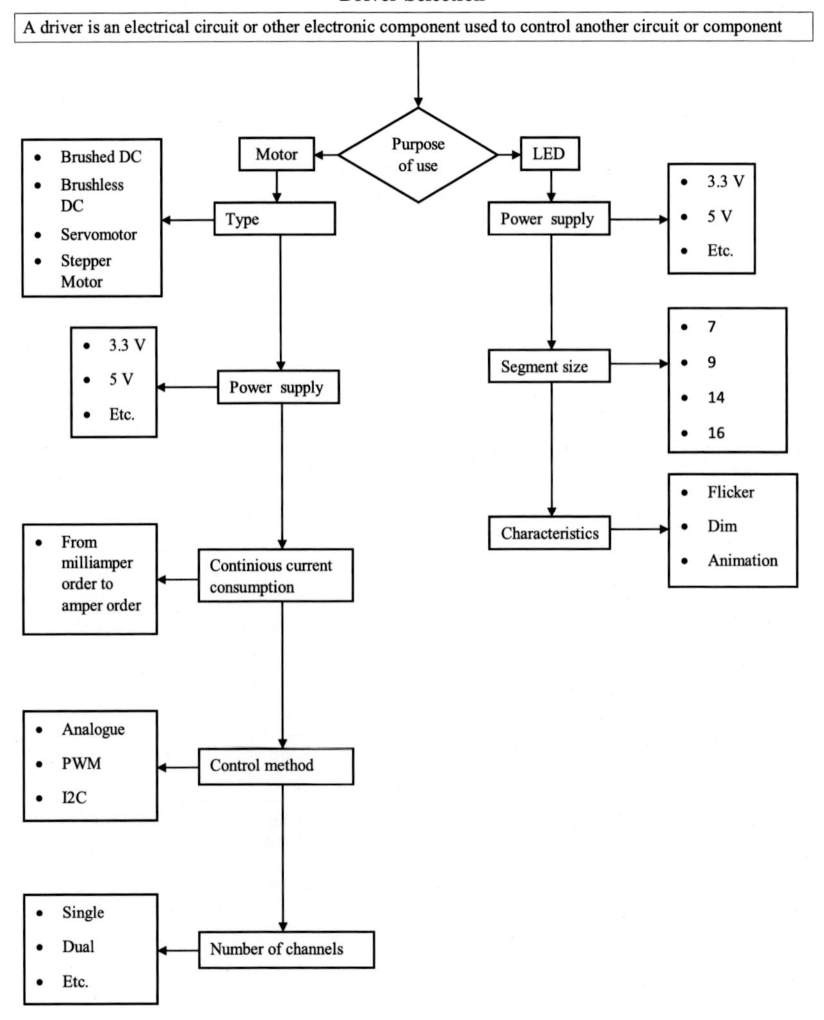
Driver Selection
A driver is an electrical circuit or other electronic component used to control another circuit or component
Purpose of use
Motor
LED
Type
Brushed DC
Brushless DC
Servomotor
Stepper Motor
Power supply
3.3 V
5 V
Etc.
Continious current consumption
From milliamper order to amper order
Control method
Analogue
PWM
I2C
Number of channels
Single
Dual
Etc.
Power supply
3.3 V
5 V
Etc.
Segment size
7
9
14
16
Characteristics
Flicker
Dim
Animation

Motor Driver Properties

After choosing a suitable motor for our engineering project, we can choose a motor driver. The reason as to why we are using a motor driver is usually the microcontrollers or other digital controllers do not provide the necessary power to drive the motors. The motor drivers can also control the motion of the motor with the help of the microcontroller. The motor drivers properties are chosen according to the motor desired.

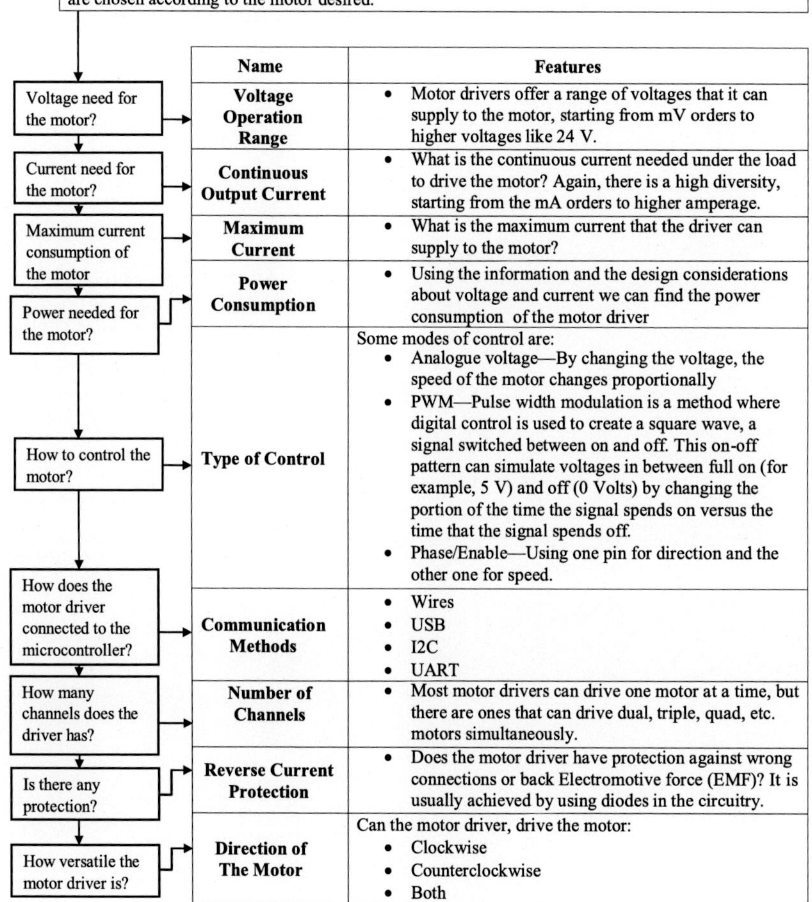

Name	Features
Voltage Operation Range	• Motor drivers offer a range of voltages that it can supply to the motor, starting from mV orders to higher voltages like 24 V.
Continuous Output Current	• What is the continuous current needed under the load to drive the motor? Again, there is a high diversity, starting from the mA orders to higher amperage.
Maximum Current	• What is the maximum current that the driver can supply to the motor?
Power Consumption	• Using the information and the design considerations about voltage and current we can find the power consumption of the motor driver
Type of Control	Some modes of control are: • Analogue voltage—By changing the voltage, the speed of the motor changes proportionally • PWM—Pulse width modulation is a method where digital control is used to create a square wave, a signal switched between on and off. This on-off pattern can simulate voltages in between full on (for example, 5 V) and off (0 Volts) by changing the portion of the time the signal spends on versus the time that the signal spends off. • Phase/Enable—Using one pin for direction and the other one for speed.
Communication Methods	• Wires • USB • I2C • UART
Number of Channels	• Most motor drivers can drive one motor at a time, but there are ones that can drive dual, triple, quad, etc. motors simultaneously.
Reverse Current Protection	• Does the motor driver have protection against wrong connections or back Electromotive force (EMF)? It is usually achieved by using diodes in the circuitry.
Direction of The Motor	Can the motor driver, drive the motor: • Clockwise • Counterclockwise • Both

Chapter 11

Actuators

ABSTRACT

Actuators are the muscle of the mechatronic systems. Mostly electrical motors are used for ease of control. DC motors are preferred in the mobile systems since the current battery technology can only store DC energy. AC motors are preferred in building since it is very easy and cheap to use electrical energy from the electrical grid. Pneumatic systems use air as an energy source while hydraulic systems use fluids, in general, oil.

Pneumatic systems are faster to respond but cannot carry high force. In contrast, the hydraulic systems can lift more forces but provide a rather slower system compared to the pneumatic systems. Since both systems are using fluids and the flow of the fluid in the piping system is nonlinear, these systems are very hard to position and speed control.

Stepper motors are good for position control while brushless DC motors are good for high-revolution needs, such as running a propeller. Regular DC motors are mostly used with gearboxes to reduce the rotation speed and to increase the torque.

Mechatronic Components. https://doi.org/10.1016/B978-0-12-814126-7.00011-6

Pneumatic Cylinders

Pneumatic cylinders, sometimes known as air cylinders, are mechanical devices which use the power of compressed gas to produce a force in a reciprocating linear motion. Like hydraulic cylinders, something forces a piston to move in the desired direction.

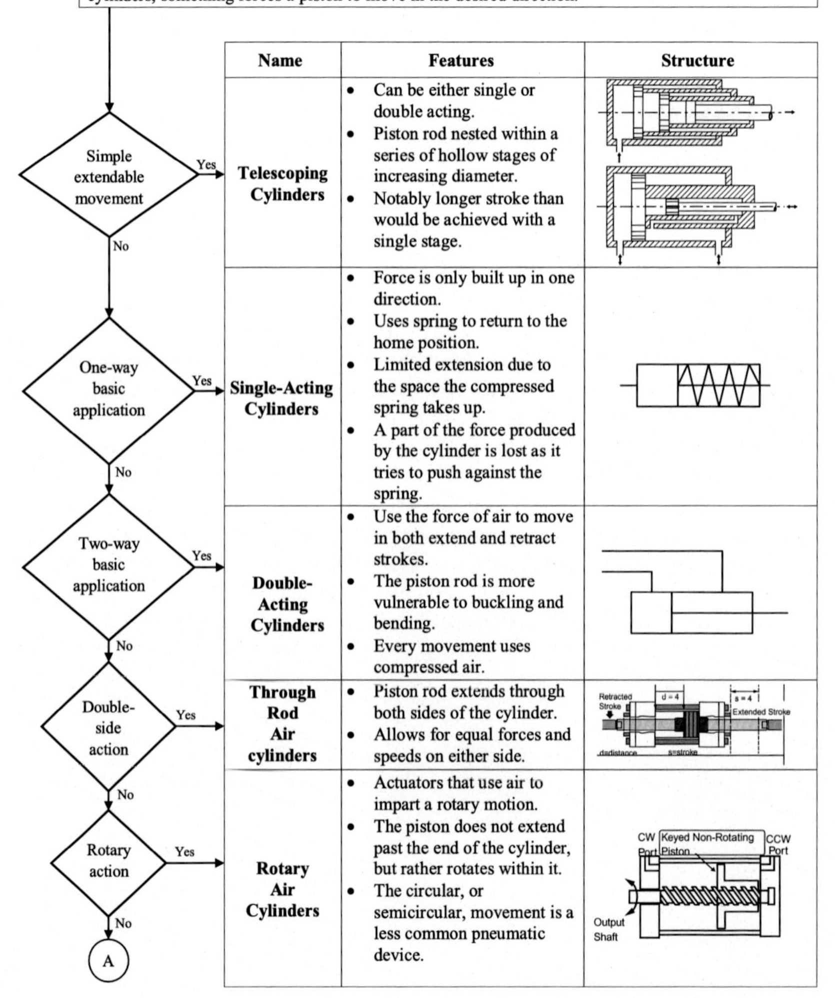

Name	Features	Structure
Telescoping Cylinders	• Can be either single or double acting. • Piston rod nested within a series of hollow stages of increasing diameter. • Notably longer stroke than would be achieved with a single stage.	
Single-Acting Cylinders	• Force is only built up in one direction. • Uses spring to return to the home position. • Limited extension due to the space the compressed spring takes up. • A part of the force produced by the cylinder is lost as it tries to push against the spring.	
Double-Acting Cylinders	• Use the force of air to move in both extend and retract strokes. • The piston rod is more vulnerable to buckling and bending. • Every movement uses compressed air.	
Through Rod Air cylinders	• Piston rod extends through both sides of the cylinder. • Allows for equal forces and speeds on either side.	
Rotary Air Cylinders	• Actuators that use air to impart a rotary motion. • The piston does not extend past the end of the cylinder, but rather rotates within it. • The circular, or semicircular, movement is a less common pneumatic device.	

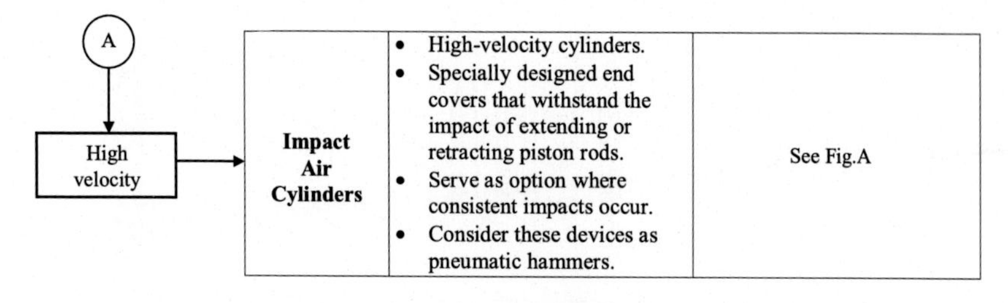

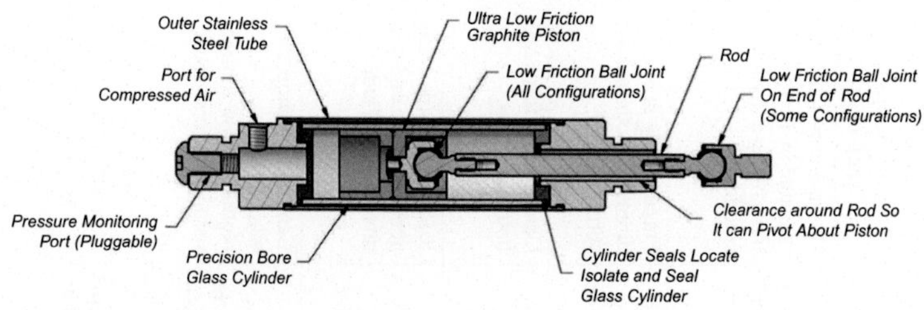

FIG. A Impact Air Cylinders Structure

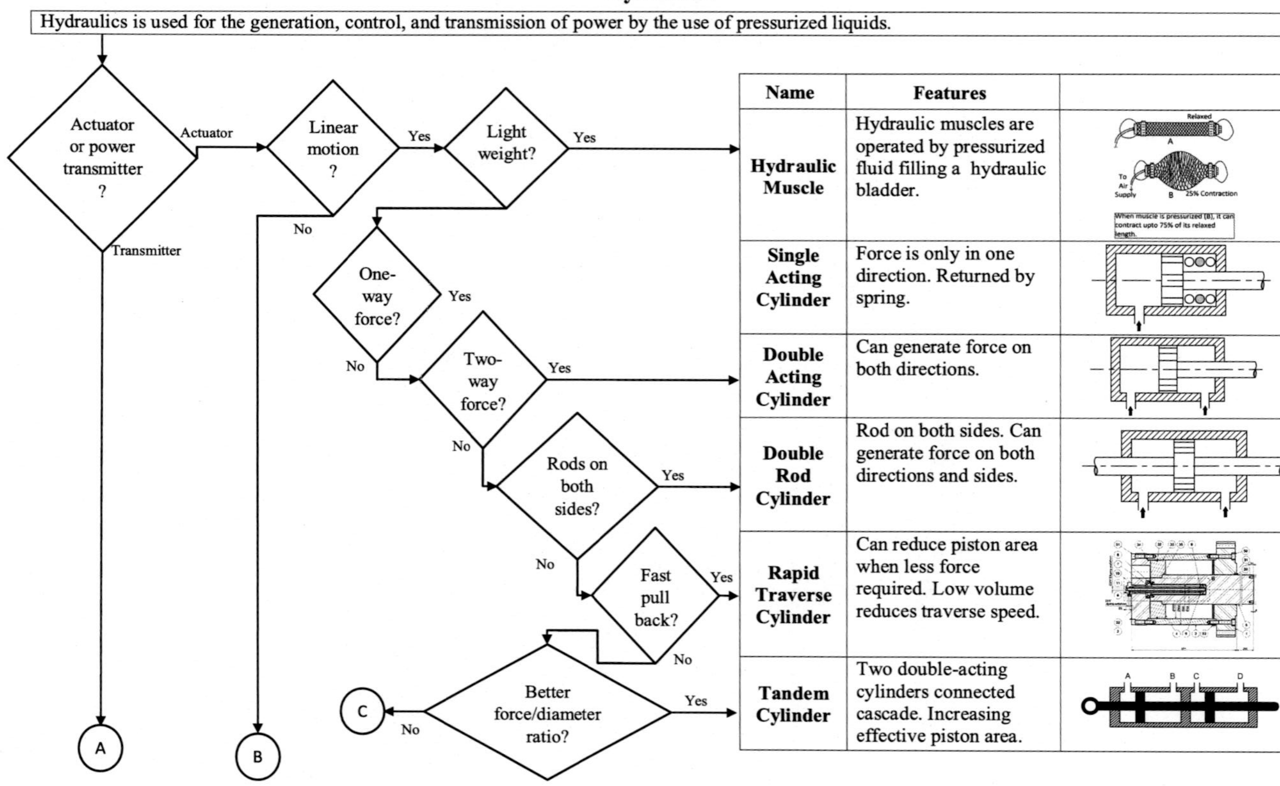
Hydraulics
Hydraulics is used for the generation, control, and transmission of power by the use of pressurized liquids.
Actuator or power transmitter ?
Actuator
Transmitter
Linear motion ?
Yes
No
Light weight?
Yes
One-way force?
Yes
No
Two-way force?
Yes
No
Rods on both sides?
Yes
No
Fast pull back?
Yes
No
Better force/diameter ratio?
Yes
No
A
B
C
Name
Features
Hydraulic Muscle
Hydraulic muscles are operated by pressurized fluid filling a hydraulic bladder.
Single Acting Cylinder
Force is only in one direction. Returned by spring.
Double Acting Cylinder
Can generate force on both directions.
Double Rod Cylinder
Rod on both sides. Can generate force on both directions and sides.
Rapid Traverse Cylinder
Can reduce piston area when less force required. Low volume reduces traverse speed.
Tandem Cylinder
Two double-acting cylinders connected cascade. Increasing effective piston area.
Relaxed
A
To Air Supply
B
25% Contraction
When muscle is pressurized (B), it can contract upto 75% of its relaxed length.
A
B
C
D

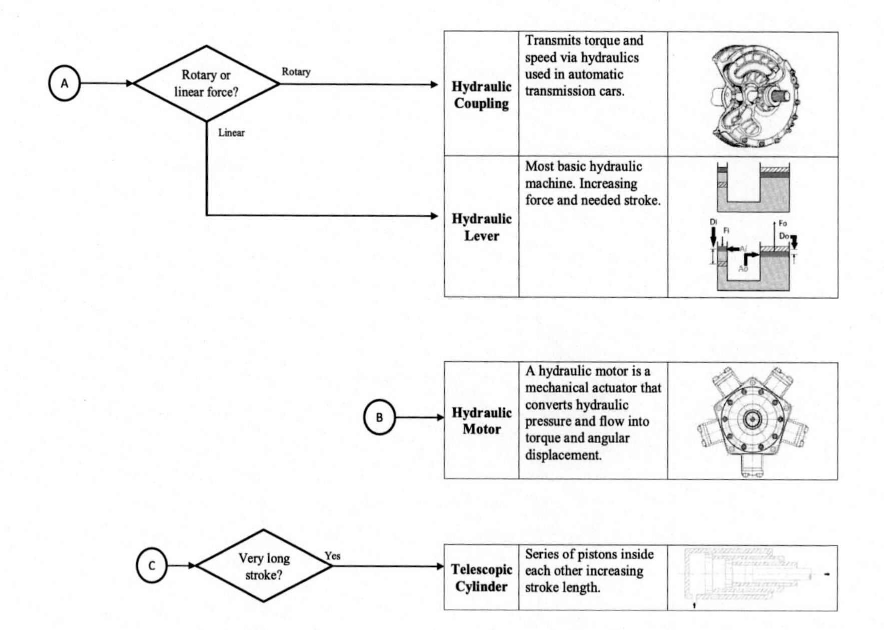
A
Rotary or linear force?
Rotary
Linear
Hydraulic Coupling
Transmits torque and speed via hydraulics used in automatic transmission cars.
Hydraulic Lever
Most basic hydraulic machine. Increasing force and needed stroke.
B
Hydraulic Motor
A hydraulic motor is a mechanical actuator that converts hydraulic pressure and flow into torque and angular displacement.
C
Very long stroke?
Yes
Telescopic Cylinder
Series of pistons inside each other increasing stroke length.

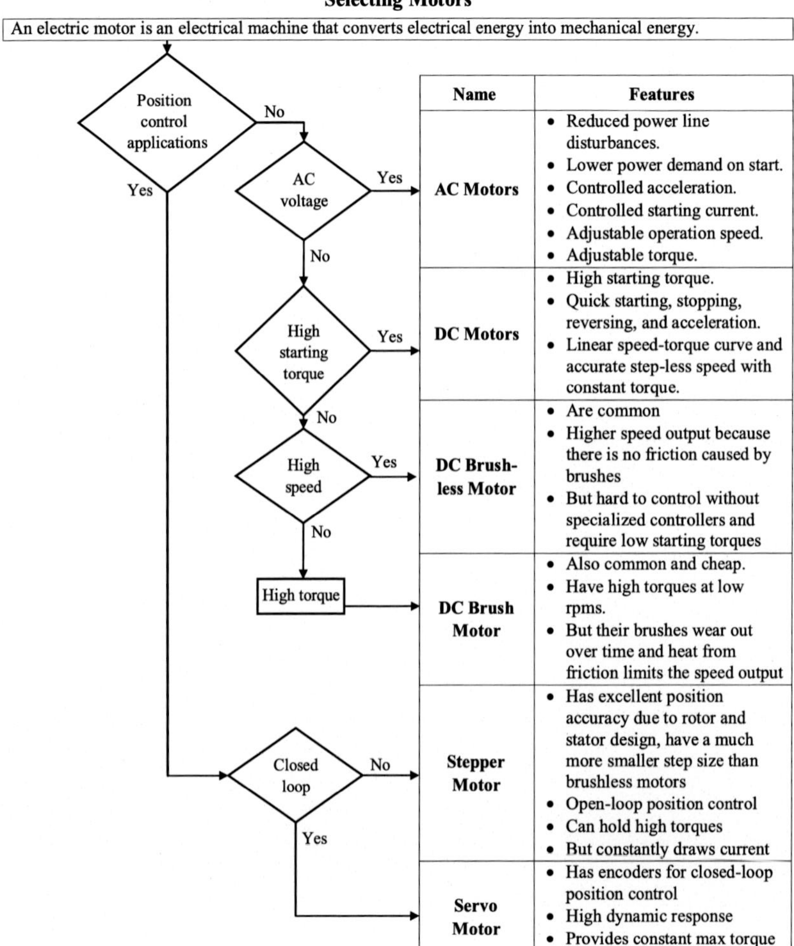

Name	Features
AC Motors	• Reduced power line disturbances. • Lower power demand on start. • Controlled acceleration. • Controlled starting current. • Adjustable operation speed. • Adjustable torque.
DC Motors	• High starting torque. • Quick starting, stopping, reversing, and acceleration. • Linear speed-torque curve and accurate step-less speed with constant torque.
DC Brush-less Motor	• Are common • Higher speed output because there is no friction caused by brushes • But hard to control without specialized controllers and require low starting torques
DC Brush Motor	• Also common and cheap. • Have high torques at low rpms. • But their brushes wear out over time and heat from friction limits the speed output
Stepper Motor	• Has excellent position accuracy due to rotor and stator design, have a much more smaller step size than brushless motors • Open-loop position control • Can hold high torques • But constantly draws current
Servo Motor	• Has encoders for closed-loop position control • High dynamic response • Provides constant max torque • Lower power consumption

Stepper Motors

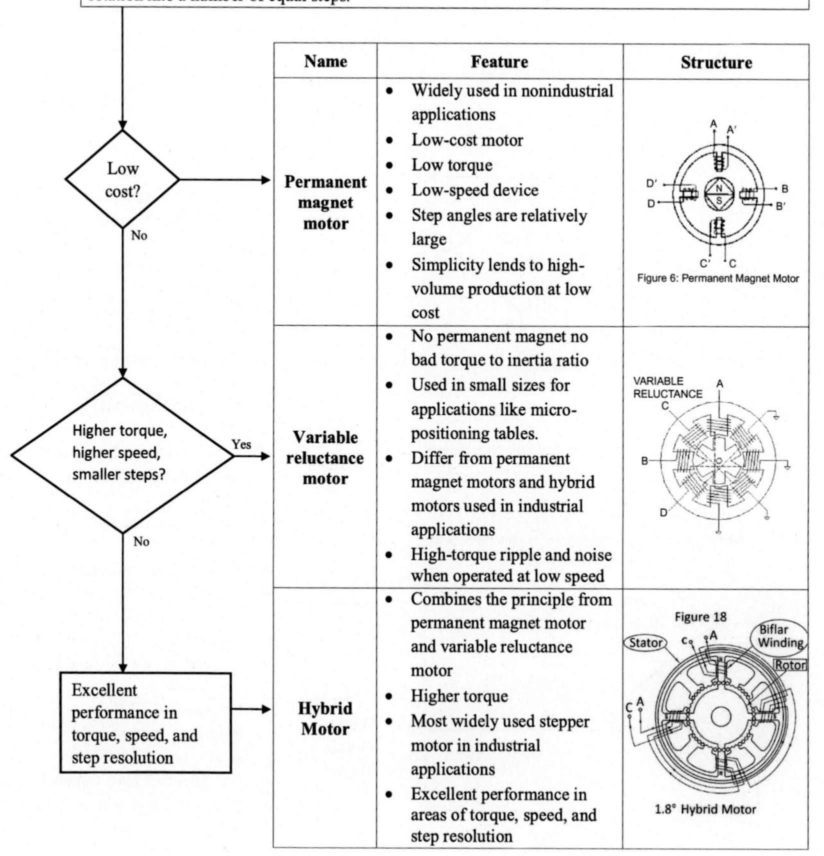

Name	Feature	Structure
Permanent magnet motor	• Widely used in nonindustrial applications • Low-cost motor • Low torque • Low-speed device • Step angles are relatively large • Simplicity lends to high-volume production at low cost	Figure 6: Permanent Magnet Motor
Variable reluctance motor	• No permanent magnet no bad torque to inertia ratio • Used in small sizes for applications like micro-positioning tables. • Differ from permanent magnet motors and hybrid motors used in industrial applications • High-torque ripple and noise when operated at low speed	
Hybrid Motor	• Combines the principle from permanent magnet motor and variable reluctance motor • Higher torque • Most widely used stepper motor in industrial applications • Excellent performance in areas of torque, speed, and step resolution	1.8° Hybrid Motor

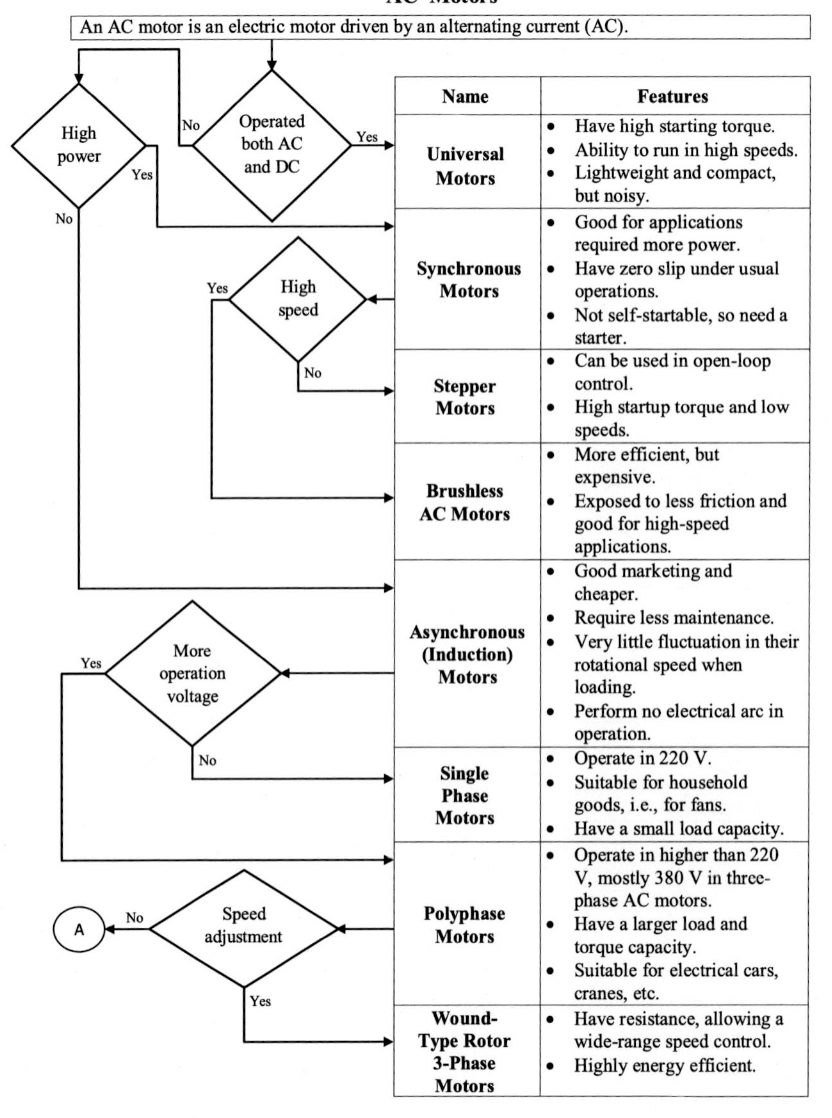

Name	Features
Universal Motors	• Have high starting torque. • Ability to run in high speeds. • Lightweight and compact, but noisy.
Synchronous Motors	• Good for applications required more power. • Have zero slip under usual operations. • Not self-startable, so need a starter.
Stepper Motors	• Can be used in open-loop control. • High startup torque and low speeds.
Brushless AC Motors	• More efficient, but expensive. • Exposed to less friction and good for high-speed applications.
Asynchronous (Induction) Motors	• Good marketing and cheaper. • Require less maintenance. • Very little fluctuation in their rotational speed when loading. • Perform no electrical arc in operation.
Single Phase Motors	• Operate in 220 V. • Suitable for household goods, i.e., for fans. • Have a small load capacity.
Polyphase Motors	• Operate in higher than 220 V, mostly 380 V in three-phase AC motors. • Have a larger load and torque capacity. • Suitable for electrical cars, cranes, etc.
Wound-Type Rotor 3-Phase Motors	• Have resistance, allowing a wide-range speed control. • Highly energy efficient.

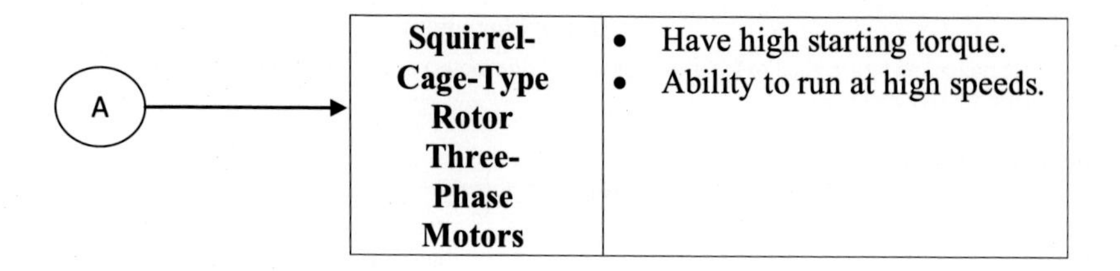
A
Squirrel-Cage-Type Rotor Three-Phase Motors
• Have high starting torque.
• Ability to run at high speeds.

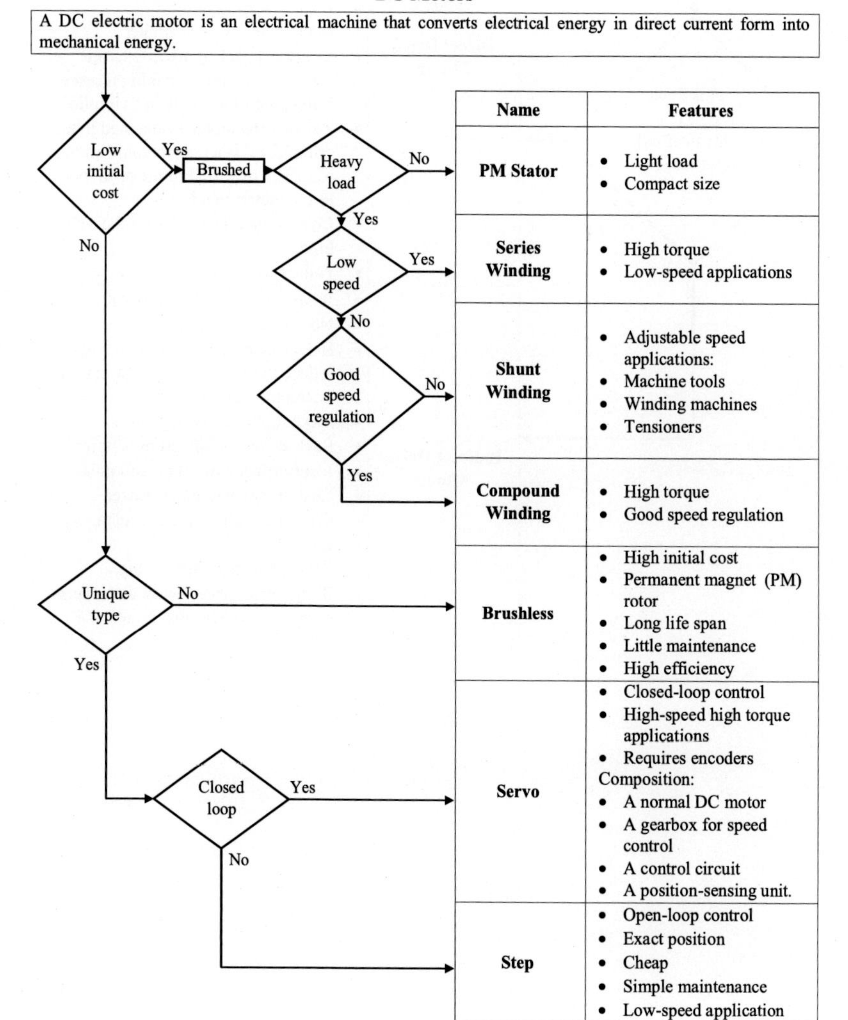
DC Motors
A DC electric motor is an electrical machine that converts electrical energy in direct current form into mechanical energy.
Low initial cost
Yes
Brushed
Heavy load
No
Yes
Low speed
Yes
No
Good speed regulation
No
Yes
No
Unique type
No
Yes
Closed loop
Yes
No
Name
Features
PM Stator
• Light load
• Compact size
Series Winding
• High torque
• Low-speed applications
Shunt Winding
• Adjustable speed applications:
• Machine tools
• Winding machines
• Tensioners
Compound Winding
• High torque
• Good speed regulation
Brushless
• High initial cost
• Permanent magnet (PM) rotor
• Long life span
• Little maintenance
• High efficiency
Servo
• Closed-loop control
• High-speed high torque applications
• Requires encoders
Composition:
• A normal DC motor
• A gearbox for speed control
• A control circuit
• A position-sensing unit.
Step
• Open-loop control
• Exact position
• Cheap
• Simple maintenance
• Low-speed application

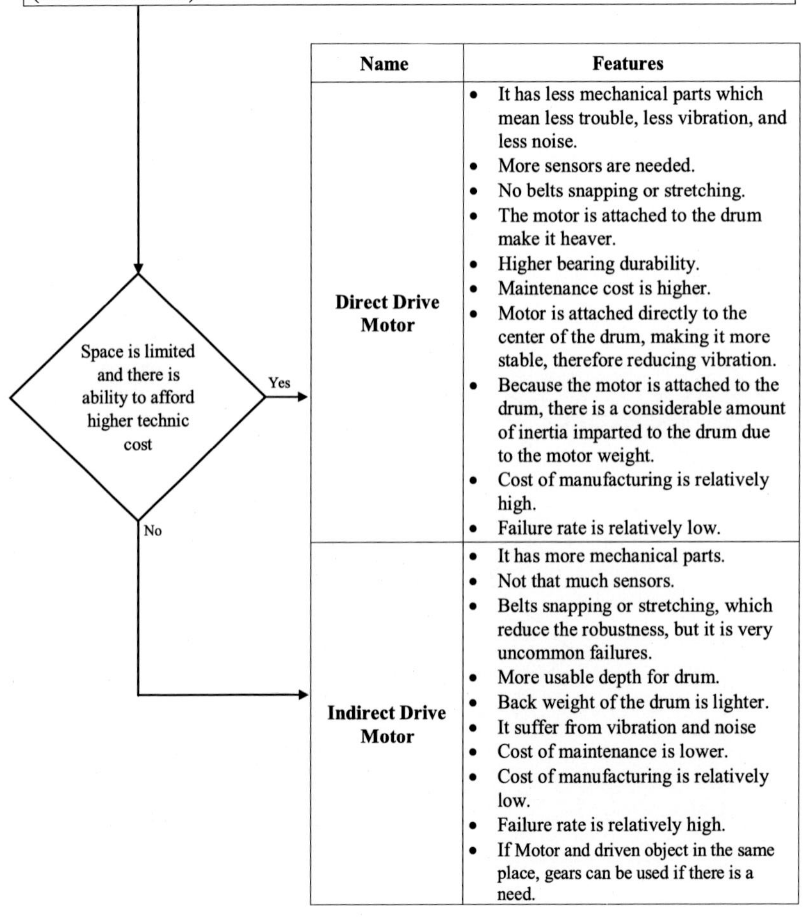

Name	Features
Direct Drive Motor	• It has less mechanical parts which mean less trouble, less vibration, and less noise. • More sensors are needed. • No belts snapping or stretching. • The motor is attached to the drum make it heaver. • Higher bearing durability. • Maintenance cost is higher. • Motor is attached directly to the center of the drum, making it more stable, therefore reducing vibration. • Because the motor is attached to the drum, there is a considerable amount of inertia imparted to the drum due to the motor weight. • Cost of manufacturing is relatively high. • Failure rate is relatively low.
Indirect Drive Motor	• It has more mechanical parts. • Not that much sensors. • Belts snapping or stretching, which reduce the robustness, but it is very uncommon failures. • More usable depth for drum. • Back weight of the drum is lighter. • It suffer from vibration and noise • Cost of maintenance is lower. • Cost of manufacturing is relatively low. • Failure rate is relatively high. • If Motor and driven object in the same place, gears can be used if there is a need.

Chapter 12

Sensors

ABSTRACT

Sensors are used in mechatronic systems to be able to measure the system and use these data in the control system as a feedback. There are sensors for almost every different physical (chemical and biological too) parameter such as rotation, velocity, or temperature. Even it is possible to measure some of the things that a human body cannot measure (but can understand the change), such as humidity or acceleration.

When the designer is selecting a sensor, working conditions, power source, accuracy, repeatability, and response time should be taken into account. The datasheet of a sensor is an excellent source of information and if the datasheet of the sensor is not providing the necessary information, that sensor should not be used.

Mechatronic Components. https://doi.org/10.1016/B978-0-12-814126-7.00012-8

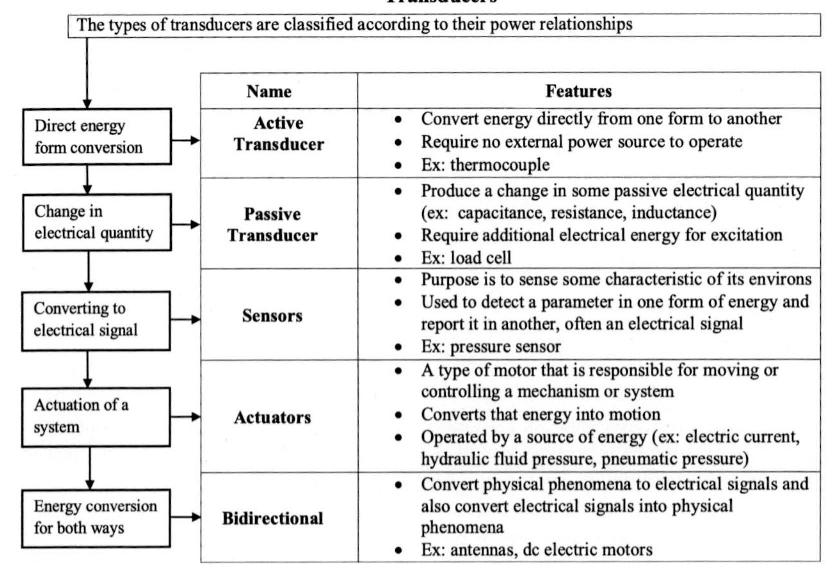

Name	Features
Active Transducer	• Convert energy directly from one form to another • Require no external power source to operate • Ex: thermocouple
Passive Transducer	• Produce a change in some passive electrical quantity (ex: capacitance, resistance, inductance) • Require additional electrical energy for excitation • Ex: load cell
Sensors	• Purpose is to sense some characteristic of its environs • Used to detect a parameter in one form of energy and report it in another, often an electrical signal • Ex: pressure sensor
Actuators	• A type of motor that is responsible for moving or controlling a mechanism or system • Converts that energy into motion • Operated by a source of energy (ex: electric current, hydraulic fluid pressure, pneumatic pressure)
Bidirectional	• Convert physical phenomena to electrical signals and also convert electrical signals into physical phenomena • Ex: antennas, dc electric motors

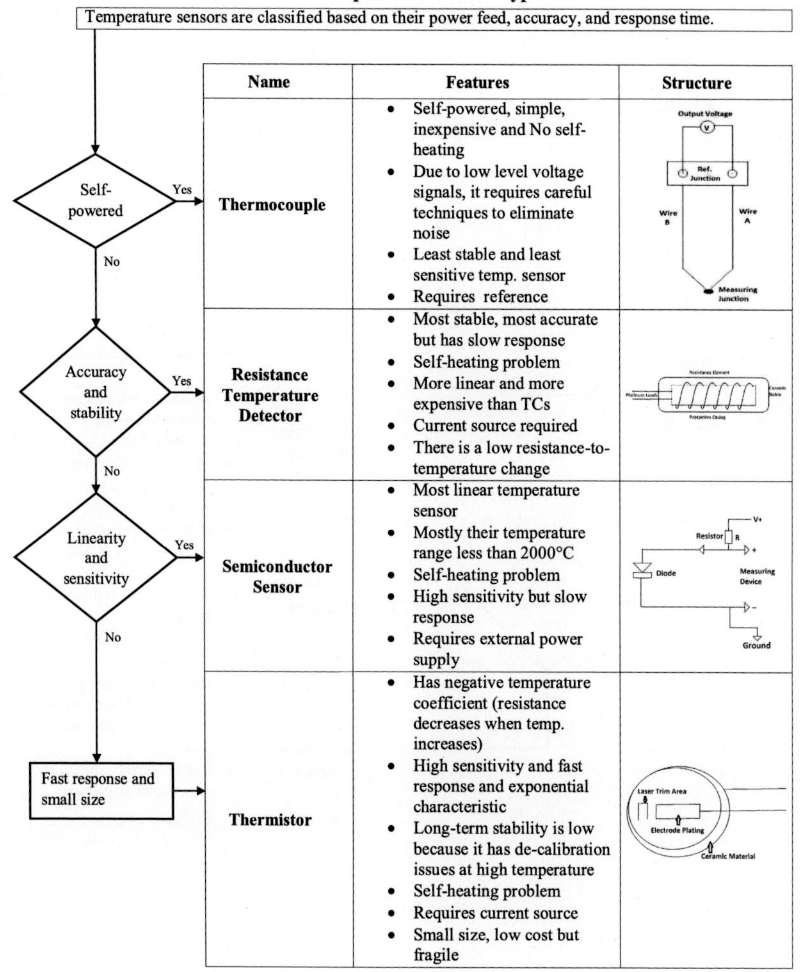

Temperature Sensor Types

Temperature sensors are classified based on their power feed, accuracy, and response time.

Name	Features	Structure
Thermocouple	• Self-powered, simple, inexpensive and No self-heating • Due to low level voltage signals, it requires careful techniques to eliminate noise • Least stable and least sensitive temp. sensor • Requires reference	
Resistance Temperature Detector	• Most stable, most accurate but has slow response • Self-heating problem • More linear and more expensive than TCs • Current source required • There is a low resistance-to-temperature change	
Semiconductor Sensor	• Most linear temperature sensor • Mostly their temperature range less than 2000°C • Self-heating problem • High sensitivity but slow response • Requires external power supply	
Thermistor	• Has negative temperature coefficient (resistance decreases when temp. increases) • High sensitivity and fast response and exponential characteristic • Long-term stability is low because it has de-calibration issues at high temperature • Self-heating problem • Requires current source • Small size, low cost but fragile	

Pressure Sensors

A pressure sensor measures pressure, typically of gases or liquids. Pressure is an expression of the force required to stop a fluid from expanding

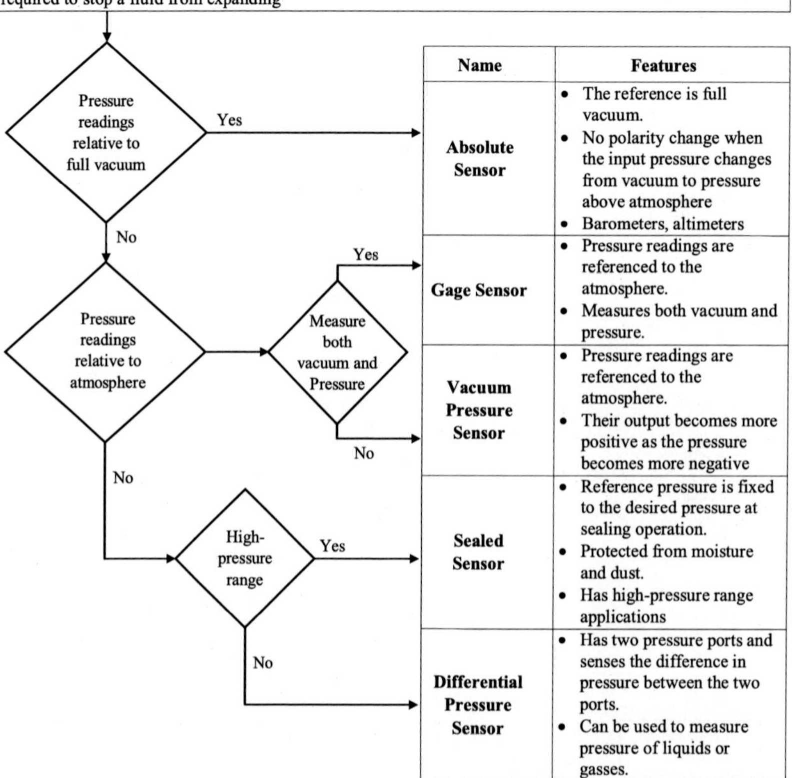

Name	Features
Absolute Sensor	• The reference is full vacuum. • No polarity change when the input pressure changes from vacuum to pressure above atmosphere • Barometers, altimeters
Gage Sensor	• Pressure readings are referenced to the atmosphere. • Measures both vacuum and pressure.
Vacuum Pressure Sensor	• Pressure readings are referenced to the atmosphere. • Their output becomes more positive as the pressure becomes more negative
Sealed Sensor	• Reference pressure is fixed to the desired pressure at sealing operation. • Protected from moisture and dust. • Has high-pressure range applications
Differential Pressure Sensor	• Has two pressure ports and senses the difference in pressure between the two ports. • Can be used to measure pressure of liquids or gasses.

Humidity Sensors

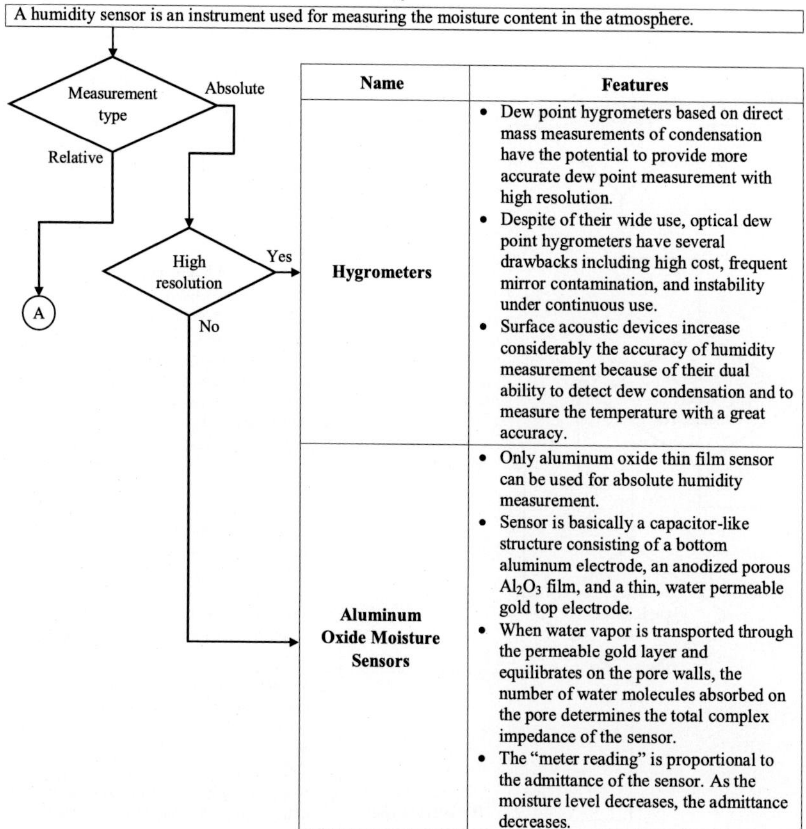

Name	Features
Hygrometers	• Dew point hygrometers based on direct mass measurements of condensation have the potential to provide more accurate dew point measurement with high resolution. • Despite of their wide use, optical dew point hygrometers have several drawbacks including high cost, frequent mirror contamination, and instability under continuous use. • Surface acoustic devices increase considerably the accuracy of humidity measurement because of their dual ability to detect dew condensation and to measure the temperature with a great accuracy.
Aluminum Oxide Moisture Sensors	• Only aluminum oxide thin film sensor can be used for absolute humidity measurement. • Sensor is basically a capacitor-like structure consisting of a bottom aluminum electrode, an anodized porous Al_2O_3 film, and a thin, water permeable gold top electrode. • When water vapor is transported through the permeable gold layer and equilibrates on the pore walls, the number of water molecules absorbed on the pore determines the total complex impedance of the sensor. • The "meter reading" is proportional to the admittance of the sensor. As the moisture level decreases, the admittance decreases.

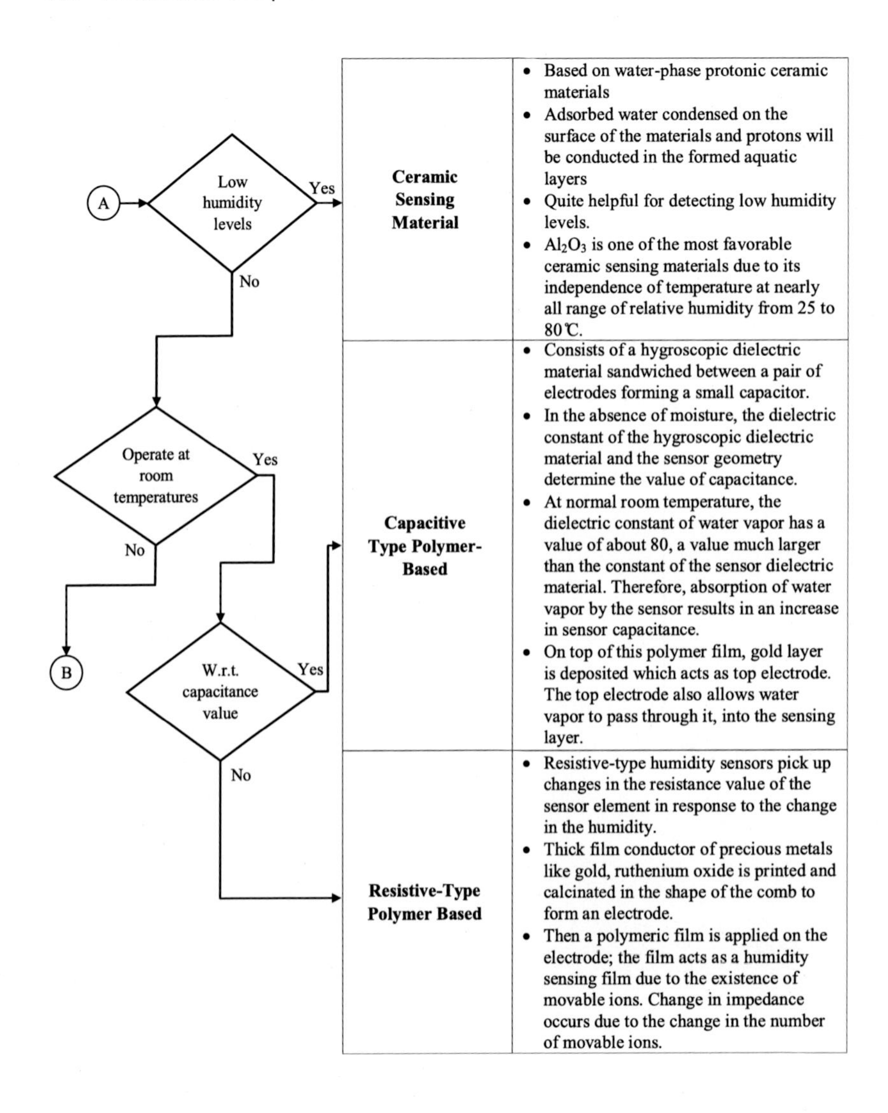

Ceramic Sensing Material	• Based on water-phase protonic ceramic materials • Adsorbed water condensed on the surface of the materials and protons will be conducted in the formed aquatic layers • Quite helpful for detecting low humidity levels. • Al_2O_3 is one of the most favorable ceramic sensing materials due to its independence of temperature at nearly all range of relative humidity from 25 to 80℃.
Capacitive Type Polymer-Based	• Consists of a hygroscopic dielectric material sandwiched between a pair of electrodes forming a small capacitor. • In the absence of moisture, the dielectric constant of the hygroscopic dielectric material and the sensor geometry determine the value of capacitance. • At normal room temperature, the dielectric constant of water vapor has a value of about 80, a value much larger than the constant of the sensor dielectric material. Therefore, absorption of water vapor by the sensor results in an increase in sensor capacitance. • On top of this polymer film, gold layer is deposited which acts as top electrode. The top electrode also allows water vapor to pass through it, into the sensing layer.
Resistive-Type Polymer Based	• Resistive-type humidity sensors pick up changes in the resistance value of the sensor element in response to the change in the humidity. • Thick film conductor of precious metals like gold, ruthenium oxide is printed and calcinated in the shape of the comb to form an electrode. • Then a polymeric film is applied on the electrode; the film acts as a humidity sensing film due to the existence of movable ions. Change in impedance occurs due to the change in the number of movable ions.

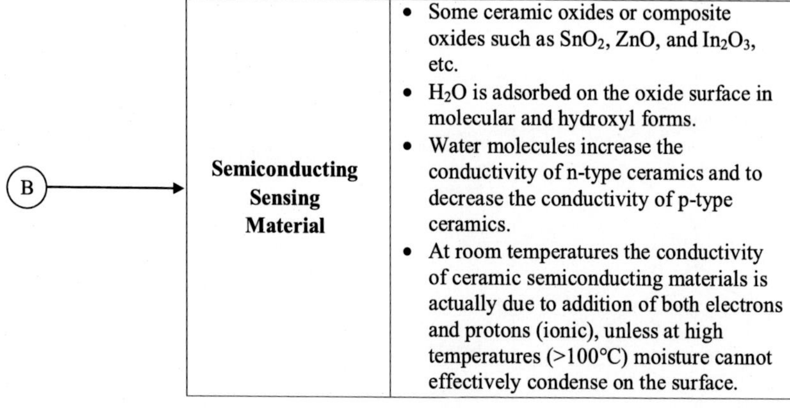

B → Semiconducting Sensing Material	
Semiconducting Sensing Material	• Some ceramic oxides or composite oxides such as SnO_2, ZnO, and In_2O_3, etc. • H_2O is adsorbed on the oxide surface in molecular and hydroxyl forms. • Water molecules increase the conductivity of n-type ceramics and to decrease the conductivity of p-type ceramics. • At room temperatures the conductivity of ceramic semiconducting materials is actually due to addition of both electrons and protons (ionic), unless at high temperatures (>100°C) moisture cannot effectively condense on the surface.

Range Sensors

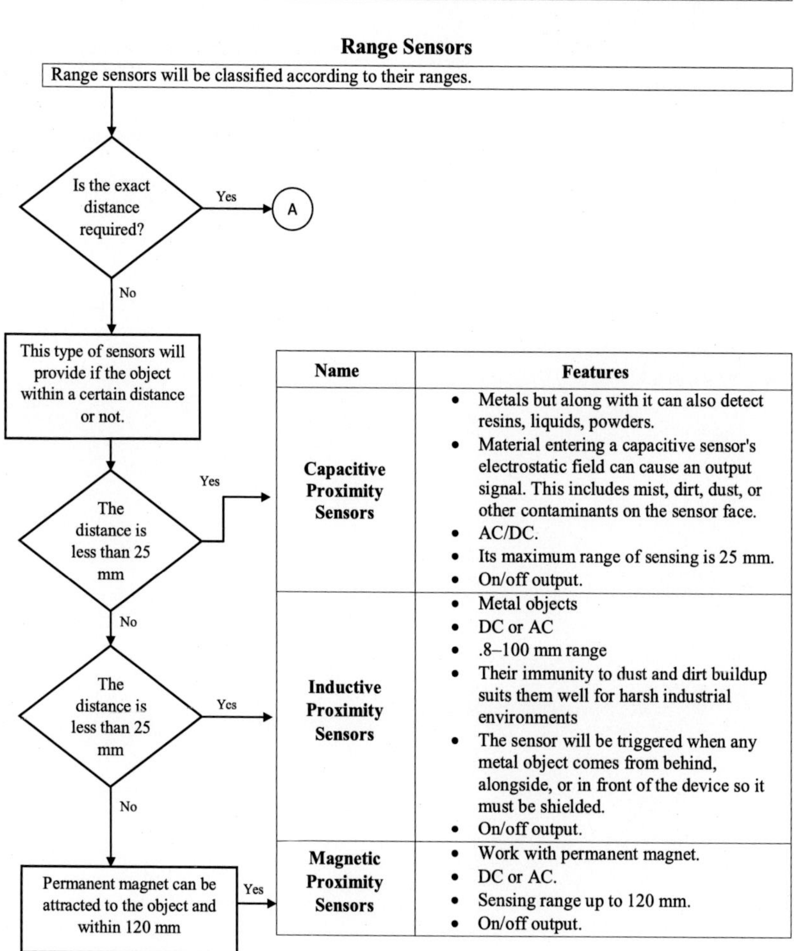

Name	Features
Capacitive Proximity Sensors	• Metals but along with it can also detect resins, liquids, powders. • Material entering a capacitive sensor's electrostatic field can cause an output signal. This includes mist, dirt, dust, or other contaminants on the sensor face. • AC/DC. • Its maximum range of sensing is 25 mm. • On/off output.
Inductive Proximity Sensors	• Metal objects • DC or AC • .8–100 mm range • Their immunity to dust and dirt buildup suits them well for harsh industrial environments • The sensor will be triggered when any metal object comes from behind, alongside, or in front of the device so it must be shielded. • On/off output.
Magnetic Proximity Sensors	• Work with permanent magnet. • DC or AC. • Sensing range up to 120 mm. • On/off output.

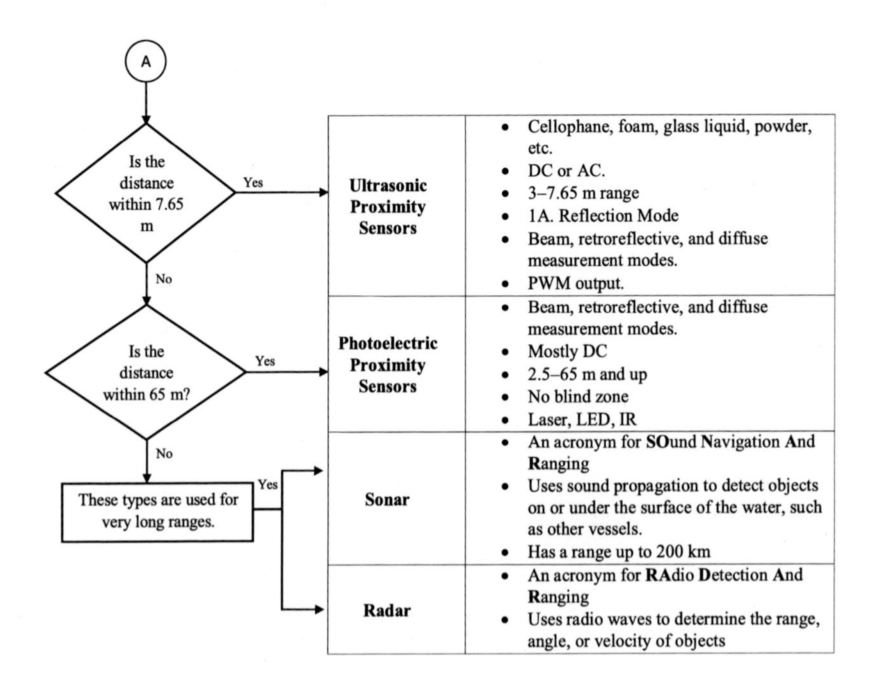
A
Is the distance within 7.65 m
Yes
No
Is the distance within 65 m?
Yes
No
These types are used for very long ranges.
Yes
Ultrasonic Proximity Sensors
Cellophane, foam, glass liquid, powder, etc.
DC or AC.
3–7.65 m range
1A. Reflection Mode
Beam, retroreflective, and diffuse measurement modes.
PWM output.
Photoelectric Proximity Sensors
Beam, retroreflective, and diffuse measurement modes.
Mostly DC
2.5–65 m and up
No blind zone
Laser, LED, IR
Sonar
An acronym for SOund Navigation And Ranging
Uses sound propagation to detect objects on or under the surface of the water, such as other vessels.
Has a range up to 200 km
Radar
An acronym for RAdio Detection And Ranging
Uses radio waves to determine the range, angle, or velocity of objects

Position Sensors

A position sensor is any device that permits position measurement. It can either be an absolute position sensor or a relative one (displacement sensor). Position sensors can be linear, angular, or multi-axis.

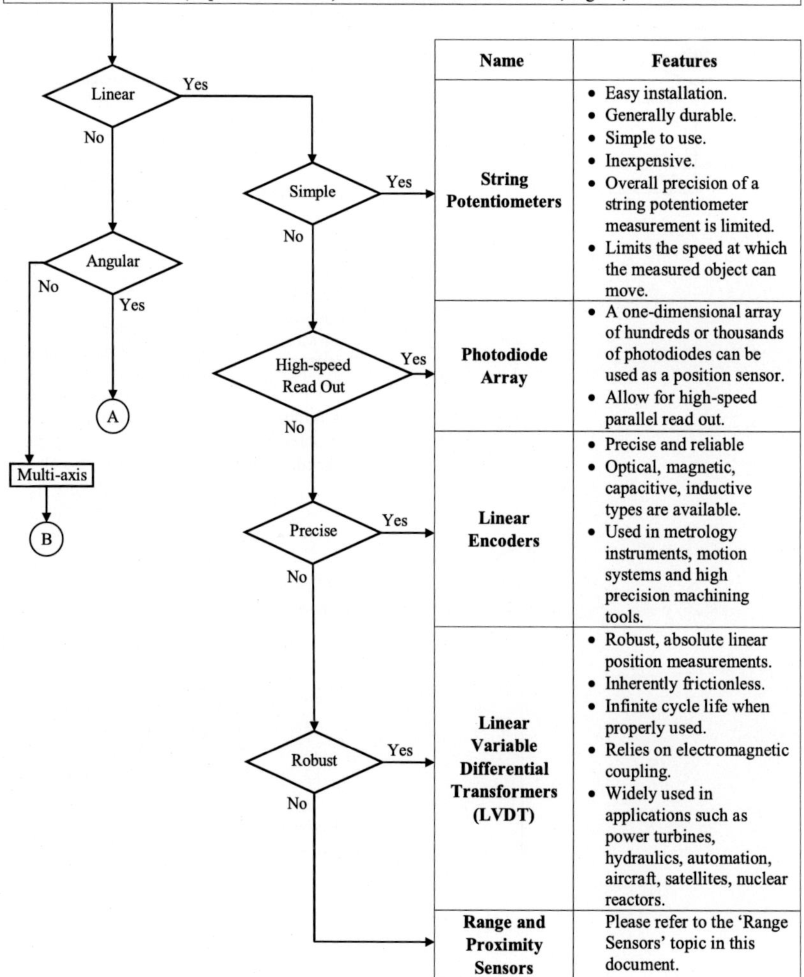

Name	Features
String Potentiometers	• Easy installation. • Generally durable. • Simple to use. • Inexpensive. • Overall precision of a string potentiometer measurement is limited. • Limits the speed at which the measured object can move.
Photodiode Array	• A one-dimensional array of hundreds or thousands of photodiodes can be used as a position sensor. • Allow for high-speed parallel read out.
Linear Encoders	• Precise and reliable • Optical, magnetic, capacitive, inductive types are available. • Used in metrology instruments, motion systems and high precision machining tools.
Linear Variable Differential Transformers (LVDT)	• Robust, absolute linear position measurements. • Inherently frictionless. • Infinite cycle life when properly used. • Relies on electromagnetic coupling. • Widely used in applications such as power turbines, hydraulics, automation, aircraft, satellites, nuclear reactors.
Range and Proximity Sensors	Please refer to the 'Range Sensors' topic in this document.

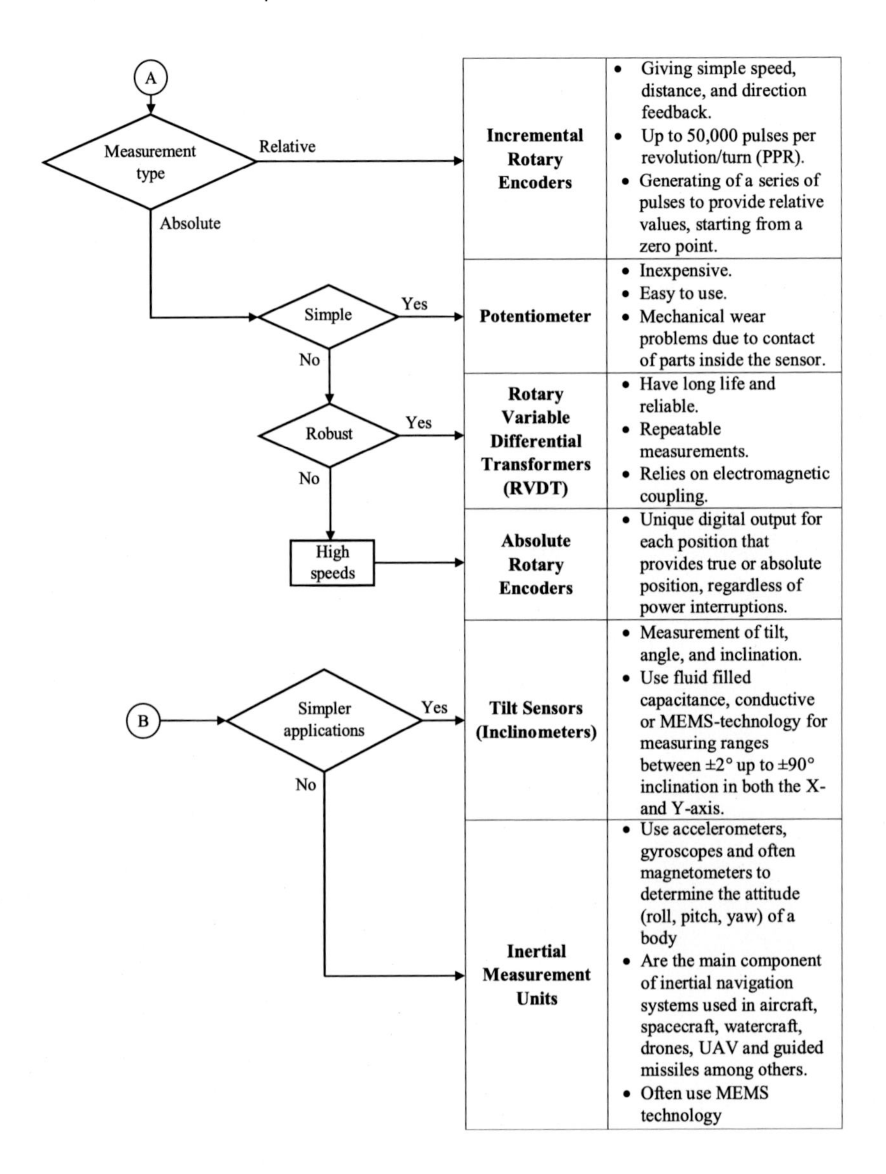

A
Measurement type
Relative
Absolute
Simple
Yes
No
Robust
Yes
No
High speeds
B
Simpler applications
Yes
No
Incremental Rotary Encoders
• Giving simple speed, distance, and direction feedback.
• Up to 50,000 pulses per revolution/turn (PPR).
• Generating of a series of pulses to provide relative values, starting from a zero point.
Potentiometer
• Inexpensive.
• Easy to use.
• Mechanical wear problems due to contact of parts inside the sensor.
Rotary Variable Differential Transformers (RVDT)
• Have long life and reliable.
• Repeatable measurements.
• Relies on electromagnetic coupling.
Absolute Rotary Encoders
• Unique digital output for each position that provides true or absolute position, regardless of power interruptions.
Tilt Sensors (Inclinometers)
• Measurement of tilt, angle, and inclination.
• Use fluid filled capacitance, conductive or MEMS-technology for measuring ranges between ±2° up to ±90° inclination in both the X- and Y-axis.
Inertial Measurement Units
• Use accelerometers, gyroscopes and often magnetometers to determine the attitude (roll, pitch, yaw) of a body
• Are the main component of inertial navigation systems used in aircraft, spacecraft, watercraft, drones, UAV and guided missiles among others.
• Often use MEMS technology

Encoders

A rotary encoder is an electromechanical device that converts the angular position or motion of a shaft or axle to an analog or digital code.

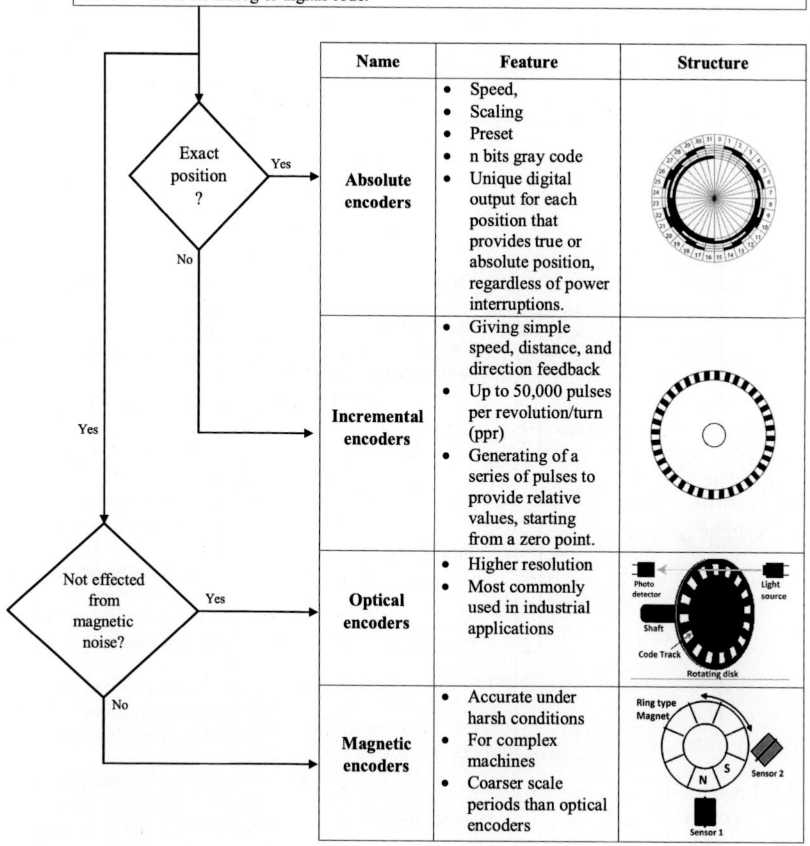

Name	Feature	Structure
Absolute encoders	• Speed, • Scaling • Preset • n bits gray code • Unique digital output for each position that provides true or absolute position, regardless of power interruptions.	
Incremental encoders	• Giving simple speed, distance, and direction feedback • Up to 50,000 pulses per revolution/turn (ppr) • Generating of a series of pulses to provide relative values, starting from a zero point.	
Optical encoders	• Higher resolution • Most commonly used in industrial applications	
Magnetic encoders	• Accurate under harsh conditions • For complex machines • Coarser scale periods than optical encoders	

Velocity Sensors

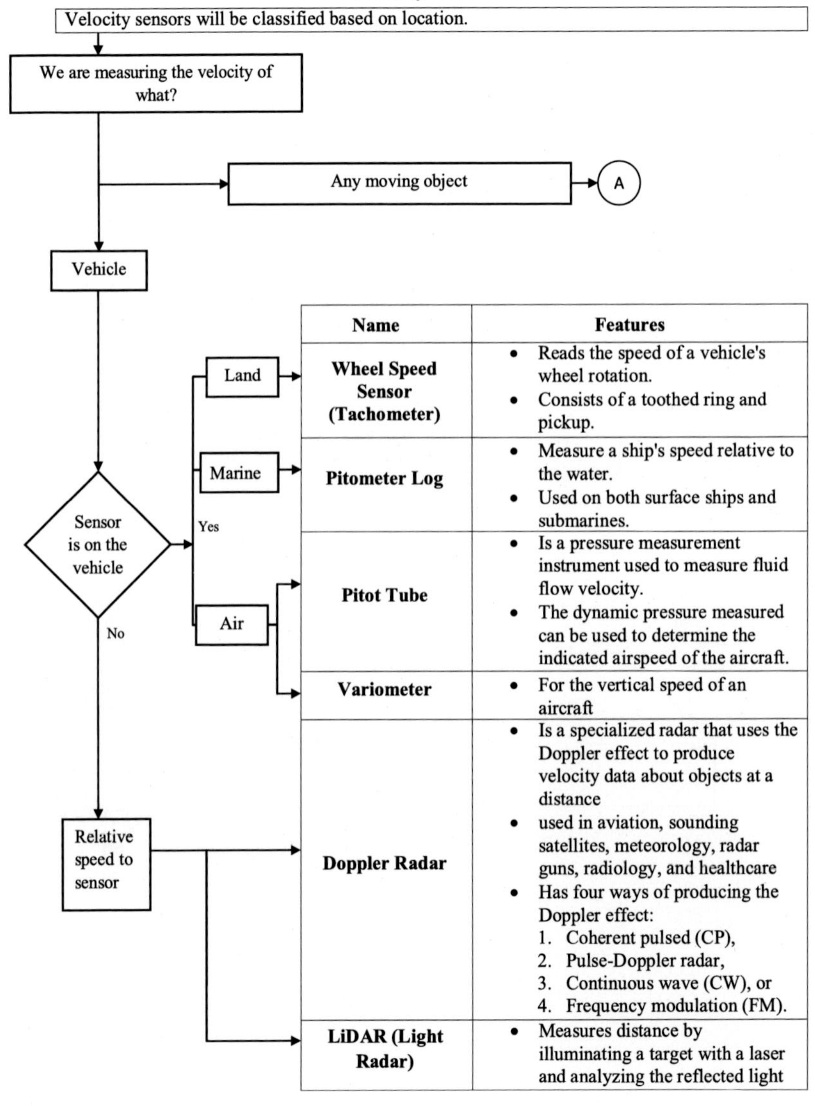

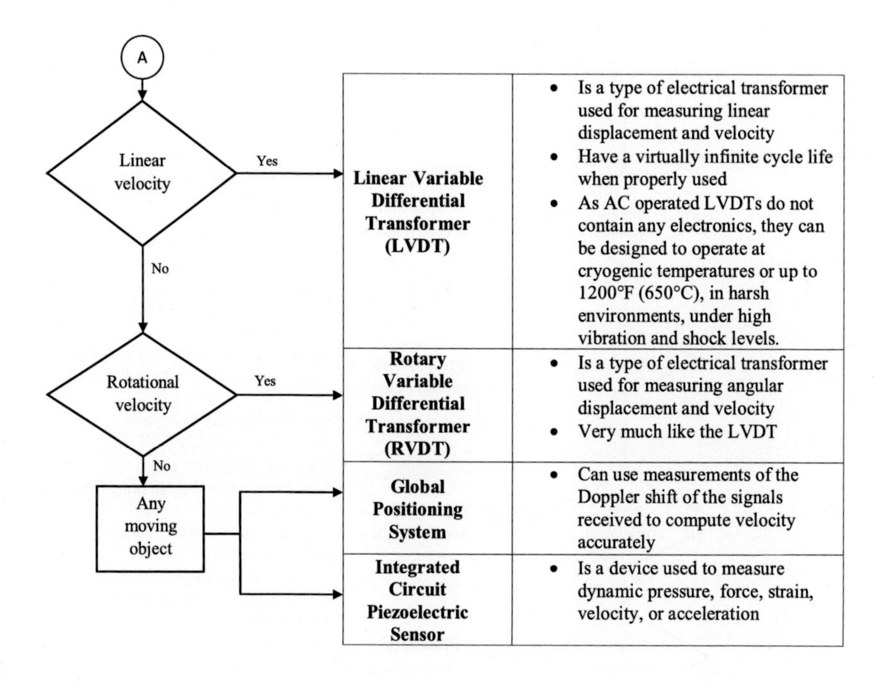

Linear Variable Differential Transformer (LVDT)	• Is a type of electrical transformer used for measuring linear displacement and velocity • Have a virtually infinite cycle life when properly used • As AC operated LVDTs do not contain any electronics, they can be designed to operate at cryogenic temperatures or up to 1200°F (650°C), in harsh environments, under high vibration and shock levels.
Rotary Variable Differential Transformer (RVDT)	• Is a type of electrical transformer used for measuring angular displacement and velocity • Very much like the LVDT
Global Positioning System	• Can use measurements of the Doppler shift of the signals received to compute velocity accurately
Integrated Circuit Piezoelectric Sensor	• Is a device used to measure dynamic pressure, force, strain, velocity, or acceleration

Accelerometer

Accelerometer is an electromechanical device that measures acceleration forces. These forces may be static or dynamic.

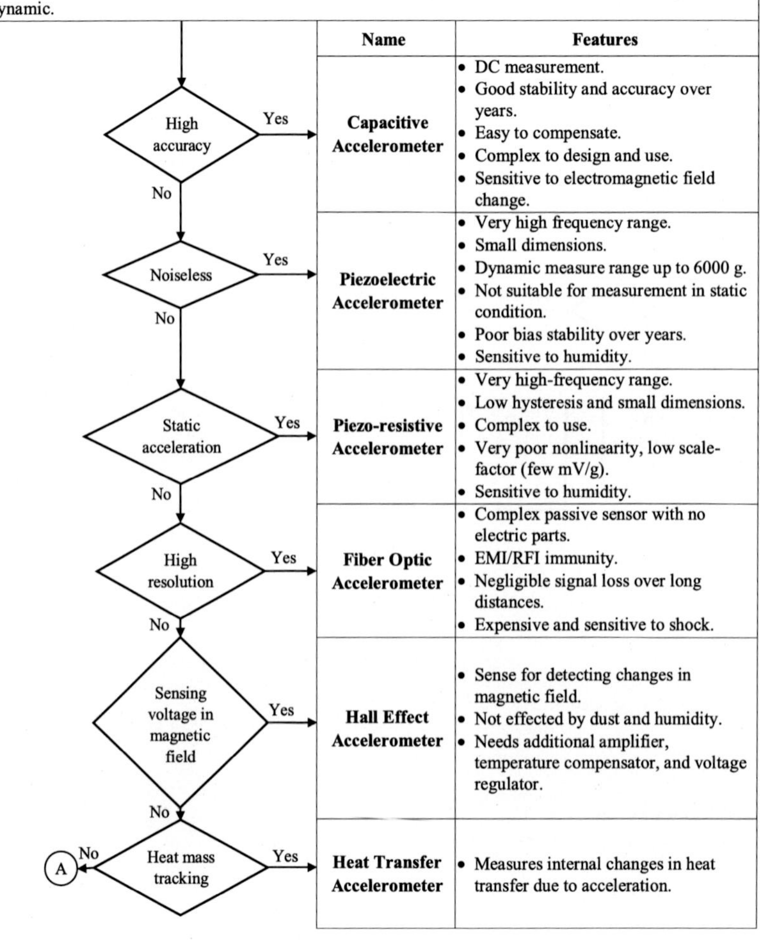

Name	Features
Capacitive Accelerometer	• DC measurement. • Good stability and accuracy over years. • Easy to compensate. • Complex to design and use. • Sensitive to electromagnetic field change.
Piezoelectric Accelerometer	• Very high frequency range. • Small dimensions. • Dynamic measure range up to 6000 g. • Not suitable for measurement in static condition. • Poor bias stability over years. • Sensitive to humidity.
Piezo-resistive Accelerometer	• Very high-frequency range. • Low hysteresis and small dimensions. • Complex to use. • Very poor nonlinearity, low scale-factor (few mV/g). • Sensitive to humidity.
Fiber Optic Accelerometer	• Complex passive sensor with no electric parts. • EMI/RFI immunity. • Negligible signal loss over long distances. • Expensive and sensitive to shock.
Hall Effect Accelerometer	• Sense for detecting changes in magnetic field. • Not effected by dust and humidity. • Needs additional amplifier, temperature compensator, and voltage regulator.
Heat Transfer Accelerometer	• Measures internal changes in heat transfer due to acceleration.

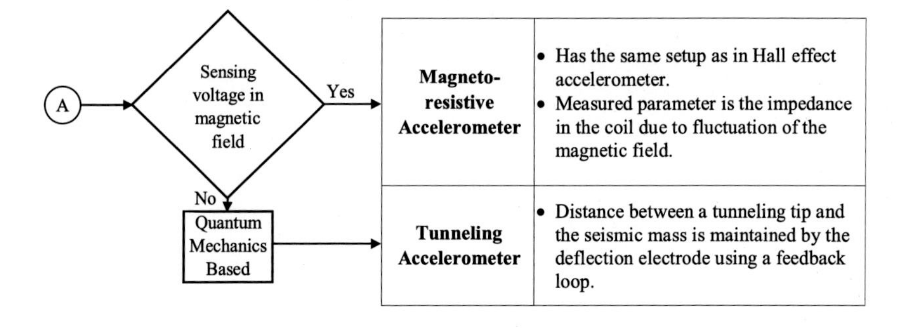

Magneto-resistive Accelerometer	• Has the same setup as in Hall effect accelerometer. • Measured parameter is the impedance in the coil due to fluctuation of the magnetic field.
Tunneling Accelerometer	• Distance between a tunneling tip and the seismic mass is maintained by the deflection electrode using a feedback loop.

Gyroscopes

A gyroscope is a spinning wheel or disc in which the axis of rotation is free to assume any orientation. When rotating, the orientation of this axis is unaffected by tilting or rotation of the mounting, according to the conservation of angular momentum. Because of this, gyroscopes are useful for measuring or maintaining orientation. Gyroscopes based on other operating principles also exist.

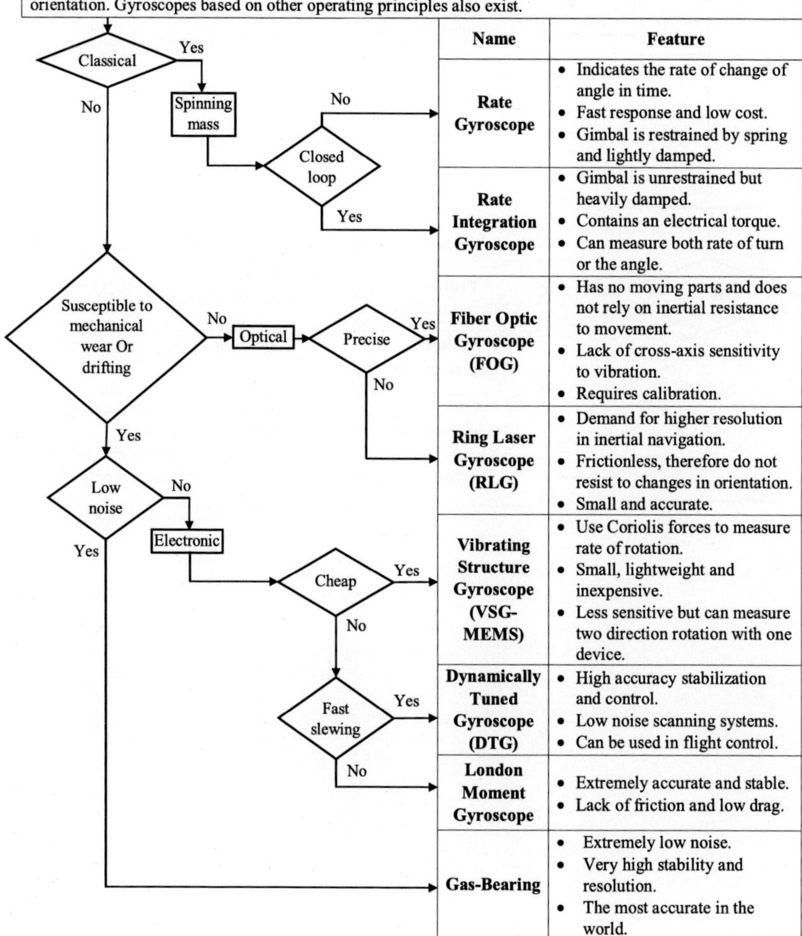

Name	Feature
Rate Gyroscope	• Indicates the rate of change of angle in time. • Fast response and low cost. • Gimbal is restrained by spring and lightly damped.
Rate Integration Gyroscope	• Gimbal is unrestrained but heavily damped. • Contains an electrical torque. • Can measure both rate of turn or the angle.
Fiber Optic Gyroscope (FOG)	• Has no moving parts and does not rely on inertial resistance to movement. • Lack of cross-axis sensitivity to vibration. • Requires calibration.
Ring Laser Gyroscope (RLG)	• Demand for higher resolution in inertial navigation. • Frictionless, therefore do not resist to changes in orientation. • Small and accurate.
Vibrating Structure Gyroscope (VSG-MEMS)	• Use Coriolis forces to measure rate of rotation. • Small, lightweight and inexpensive. • Less sensitive but can measure two direction rotation with one device.
Dynamically Tuned Gyroscope (DTG)	• High accuracy stabilization and control. • Low noise scanning systems. • Can be used in flight control.
London Moment Gyroscope	• Extremely accurate and stable. • Lack of friction and low drag.
Gas-Bearing	• Extremely low noise. • Very high stability and resolution. • The most accurate in the world.

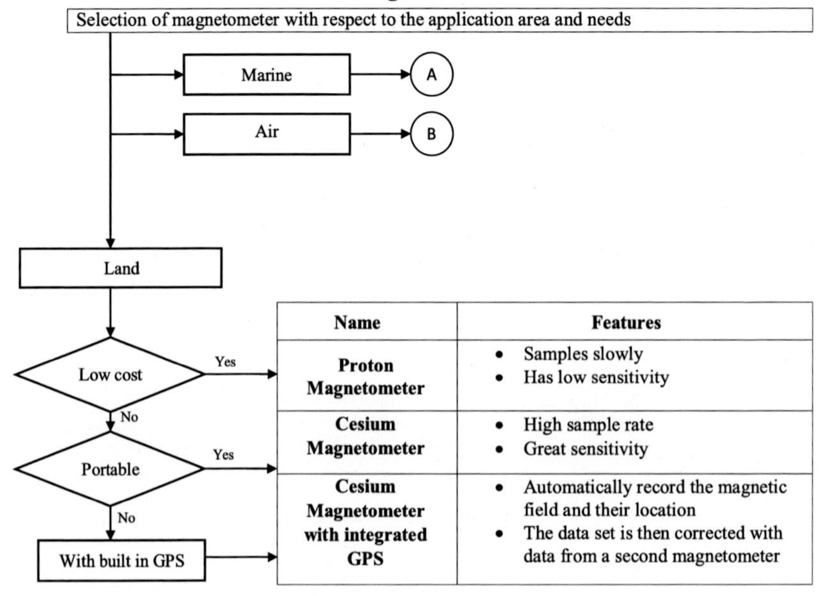

Name	Features
Proton Magnetometer	• Samples slowly • Has low sensitivity
Cesium Magnetometer	• High sample rate • Great sensitivity
Cesium Magnetometer with integrated GPS	• Automatically record the magnetic field and their location • The data set is then corrected with data from a second magnetometer

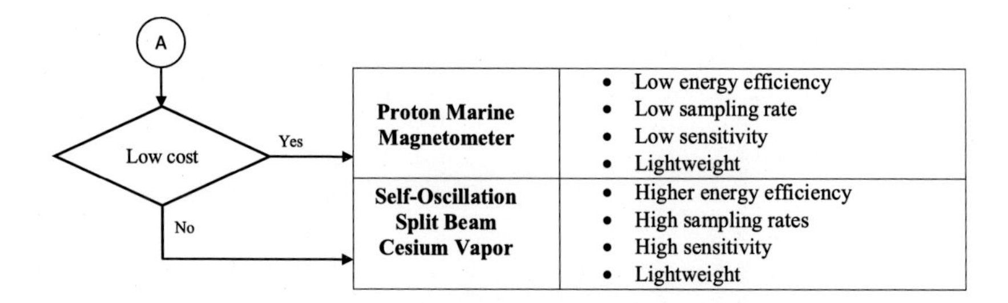
A
Low cost
Yes
No
Proton Marine Magnetometer
Low energy efficiency
Low sampling rate
Low sensitivity
Lightweight
Self-Oscillation Split Beam Cesium Vapor
Higher energy efficiency
High sampling rates
High sensitivity
Lightweight

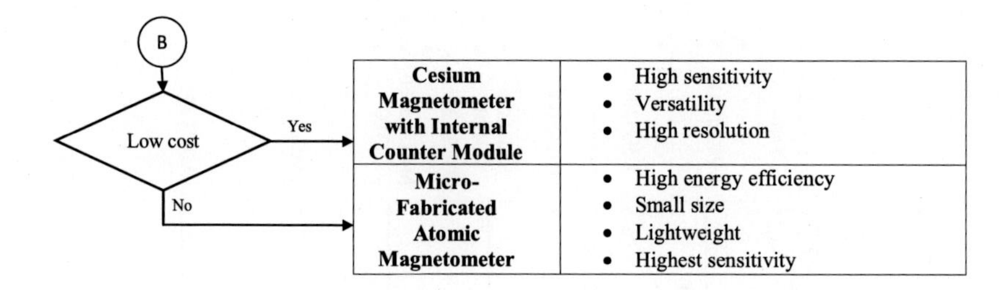
B
Low cost
Yes
No
Cesium Magnetometer with Internal Counter Module
High sensitivity
Versatility
High resolution
Micro-Fabricated Atomic Magnetometer
High energy efficiency
Small size
Lightweight
Highest sensitivity

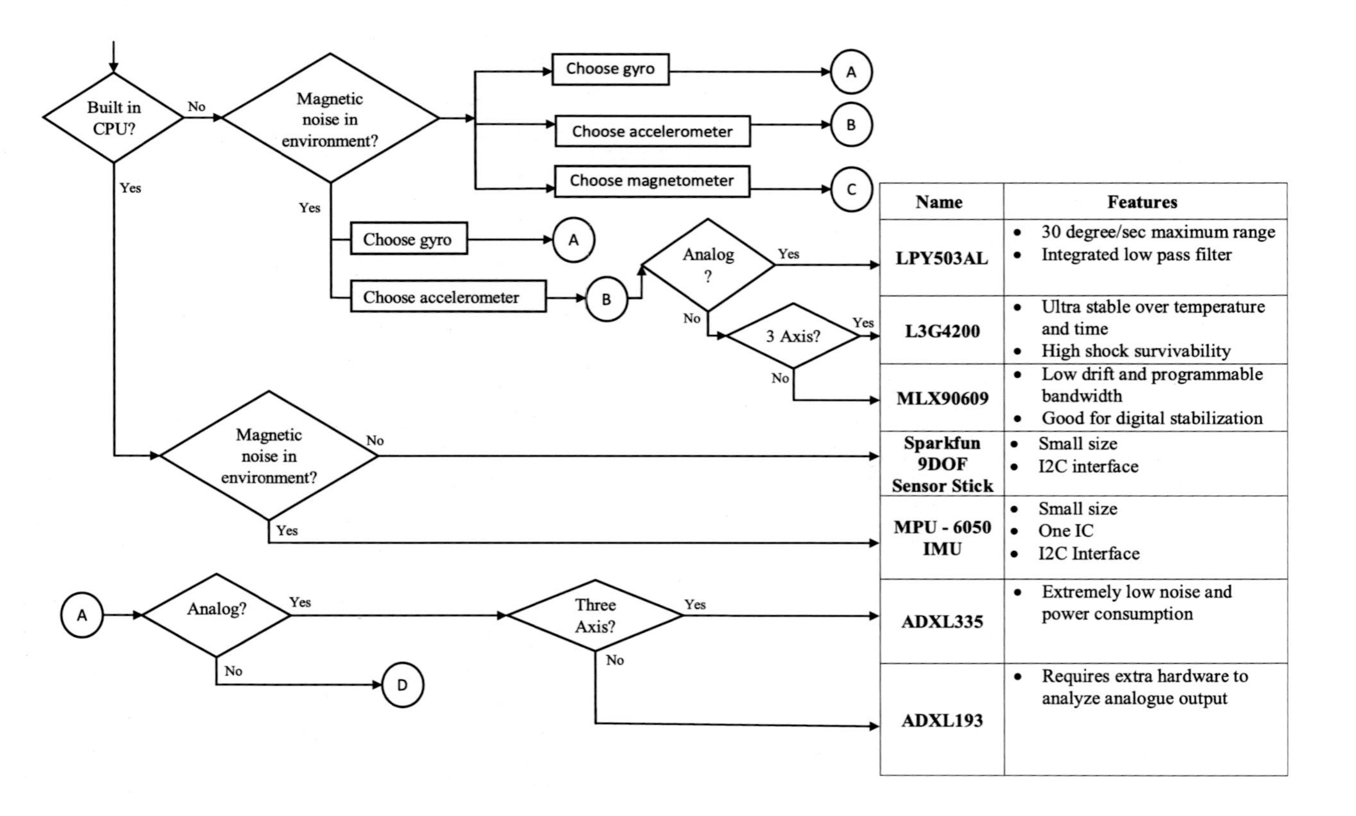

Name	Features
LPY503AL	• 30 degree/sec maximum range • Integrated low pass filter
L3G4200	• Ultra stable over temperature and time • High shock survivability
MLX90609	• Low drift and programmable bandwidth • Good for digital stabilization
Sparkfun 9DOF Sensor Stick	• Small size • I2C interface
MPU - 6050 IMU	• Small size • One IC • I2C Interface
ADXL335	• Extremely low noise and power consumption
ADXL193	• Requires extra hardware to analyze analogue output

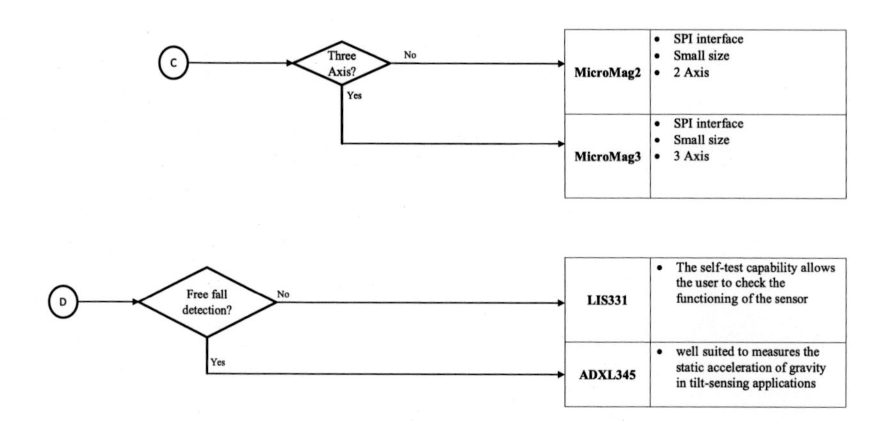
C
Three Axis?
No
Yes
MicroMag2
SPI interface
Small size
2 Axis
MicroMag3
SPI interface
Small size
3 Axis
D
Free fall detection?
No
Yes
LIS331
The self-test capability allows the user to check the functioning of the sensor
ADXL345
well suited to measures the static acceleration of gravity in tilt-sensing applications

Chapter 13

Signal Processing

ABSTRACT

In an electronic system when a signal goes around from sensor to the microcontroller or from the microcontroller to a motor driver, there is always noise in the system. A small noise will cause errors, so it should be filtered. The filters are designed depending on the signal frequency. Fourier Transform is the mathematical basis of the filtering and linearization of the signals. This methodology basically explains how any complex signal is made up from the combination of simple signals.

Mechatronic Components. https://doi.org/10.1016/B978-0-12-814126-7.00013-X

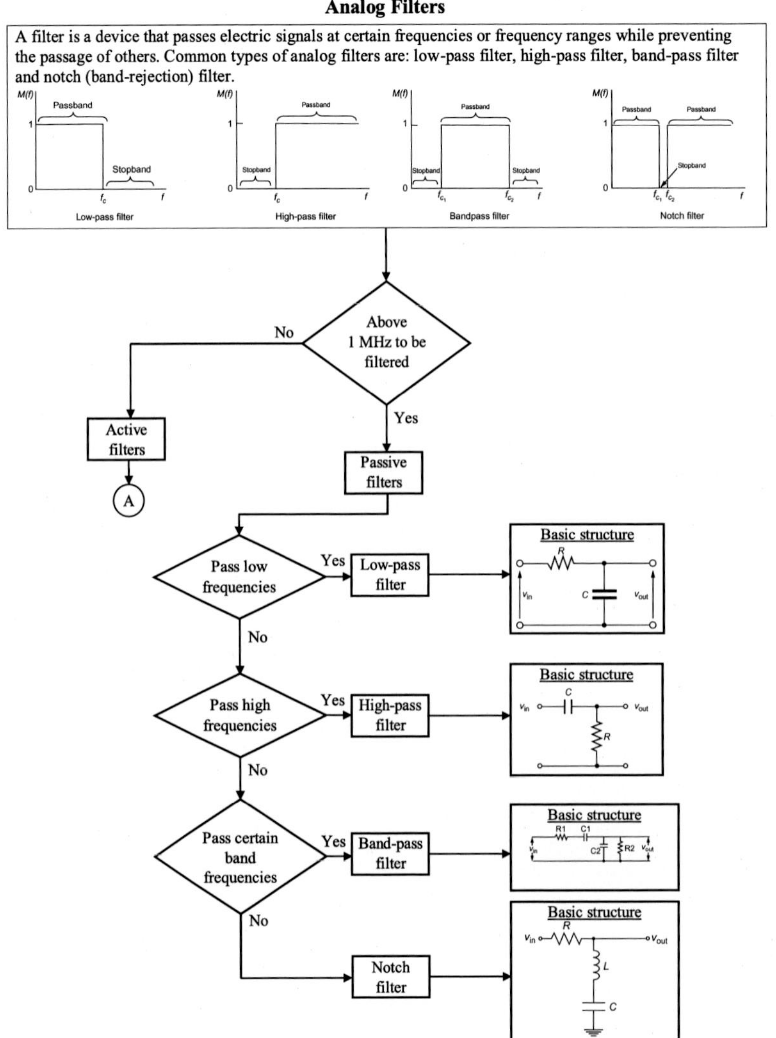
Analog Filters
A filter is a device that passes electric signals at certain frequencies or frequency ranges while preventing the passage of others. Common types of analog filters are: low-pass filter, high-pass filter, band-pass filter and notch (band-rejection) filter.
Low-pass filter
High-pass filter
Bandpass filter
Notch filter
Passband
Stopband
Above 1 MHz to be filtered
No
Yes
Active filters
A
Passive filters
Pass low frequencies
Low-pass filter
Pass high frequencies
High-pass filter
Pass certain band frequencies
Band-pass filter
Notch filter
Basic structure

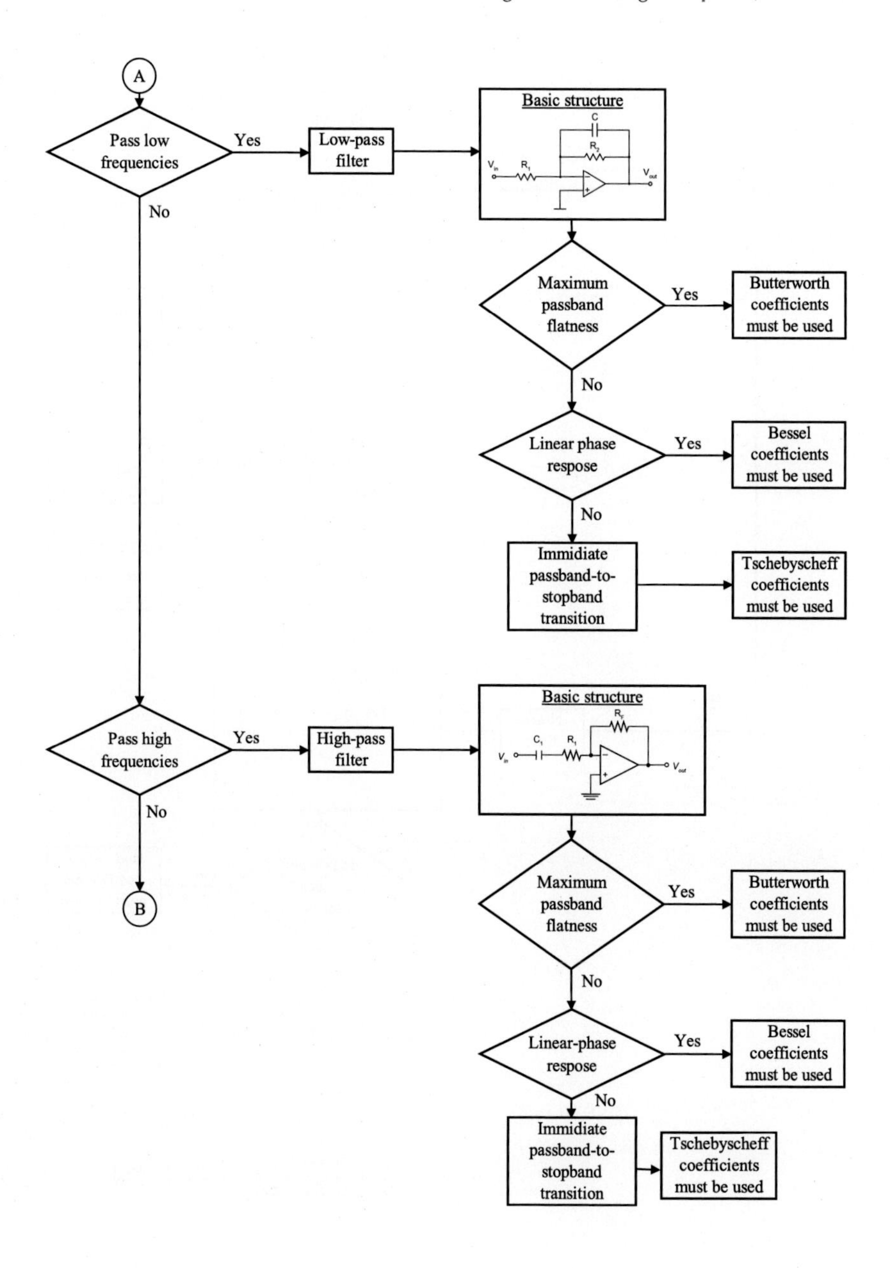
A
Pass low frequencies
Yes
No
Low-pass filter
Basic structure
C
R2
Vin
R1
Vout
Maximum passband flatness
Yes
Butterworth coefficients must be used
No
Linear phase respose
Yes
Bessel coefficients must be used
No
Immidiate passband-to-stopband transition
Tschebyscheff coefficients must be used
Pass high frequencies
Yes
No
High-pass filter
Basic structure
RF
C1
R1
Vin
Vout
B
Maximum passband flatness
Yes
Butterworth coefficients must be used
No
Linear-phase respose
Yes
Bessel coefficients must be used
No
Immidiate passband-to-stopband transition
Tschebyscheff coefficients must be used

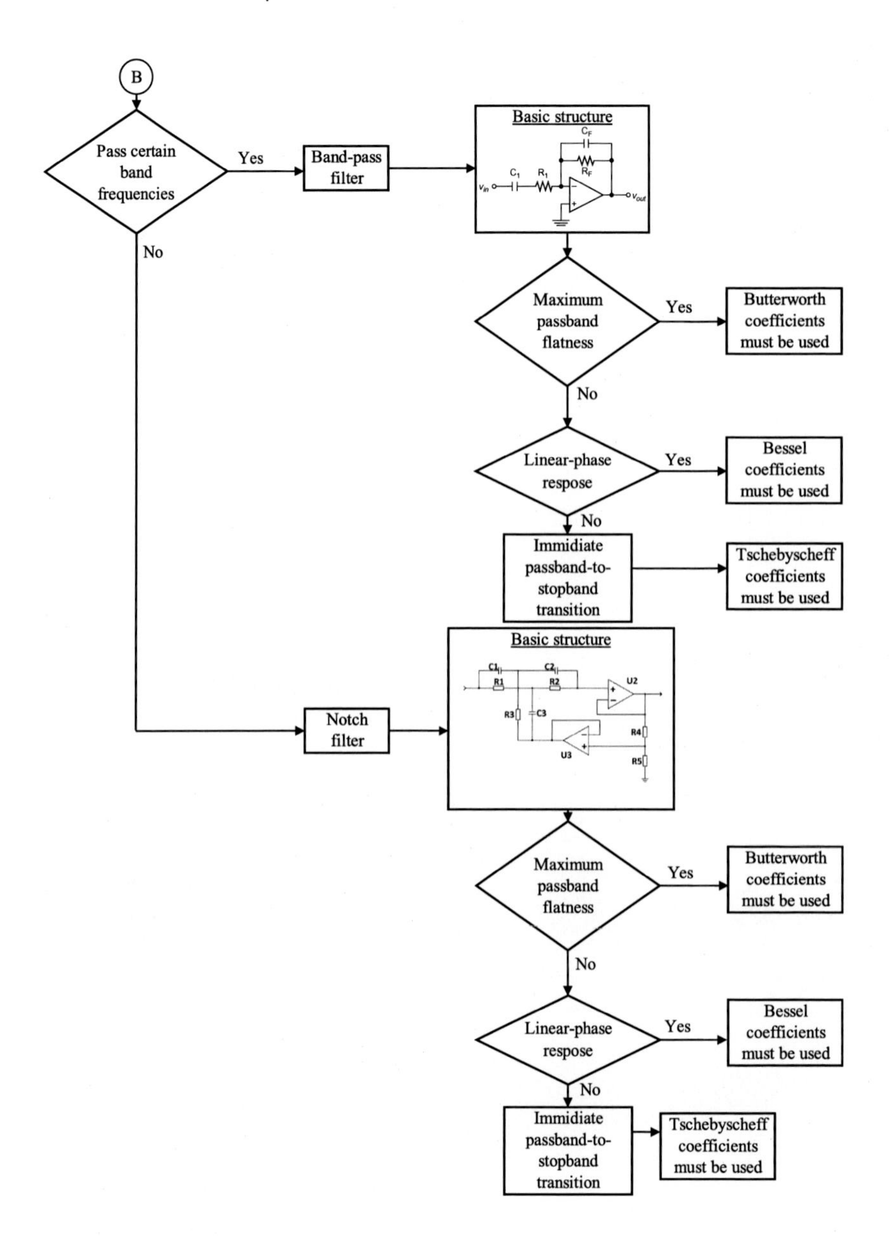
B
Pass certain band frequencies
Yes
No
Band-pass filter
Basic structure
C_F
R_F
C_1
R_1
v_in
v_out
Maximum passband flatness
Yes
Butterworth coefficients must be used
No
Linear-phase respose
Yes
Bessel coefficients must be used
No
Immidiate passband-to-stopband transition
Tschebyscheff coefficients must be used
Notch filter
Basic structure
C1
C2
R1
R2
U2
R3
C3
R4
U3
R5
Maximum passband flatness
Yes
Butterworth coefficients must be used
No
Linear-phase respose
Yes
Bessel coefficients must be used
No
Immidiate passband-to-stopband transition
Tschebyscheff coefficients must be used

Noise in Electronics

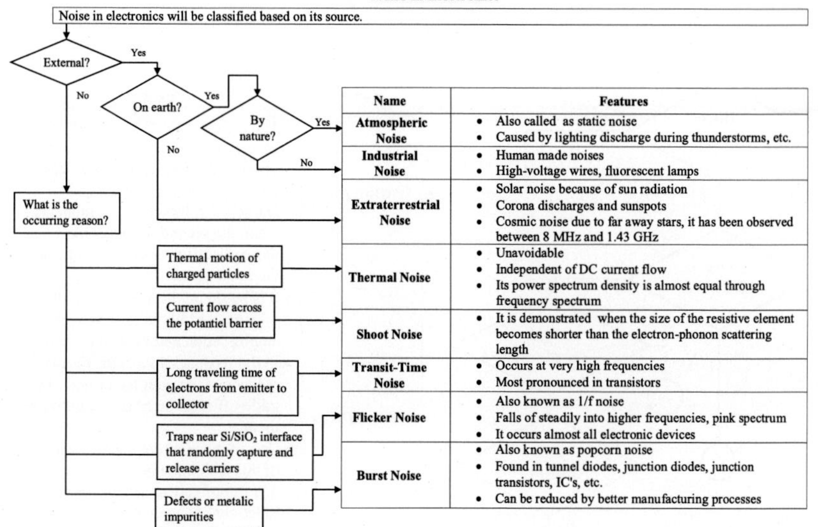

Name	Features
Atmospheric Noise	• Also called as static noise • Caused by lighting discharge during thunderstorms, etc.
Industrial Noise	• Human made noises • High-voltage wires, fluorescent lamps
Extraterrestrial Noise	• Solar noise because of sun radiation • Corona discharges and sunspots • Cosmic noise due to far away stars, it has been observed between 8 MHz and 1.43 GHz
Thermal Noise	• Unavoidable • Independent of DC current flow • Its power spectrum density is almost equal through frequency spectrum
Shoot Noise	• It is demonstrated when the size of the resistive element becomes shorter than the electron-phonon scattering length
Transit-Time Noise	• Occurs at very high frequencies • Most pronounced in transistors
Flicker Noise	• Also known as 1/f noise • Falls of steadily into higher frequencies, pink spectrum • It occurs almost all electronic devices
Burst Noise	• Also known as popcorn noise • Found in tunnel diodes, junction diodes, junction transistors, IC's, etc. • Can be reduced by better manufacturing processes

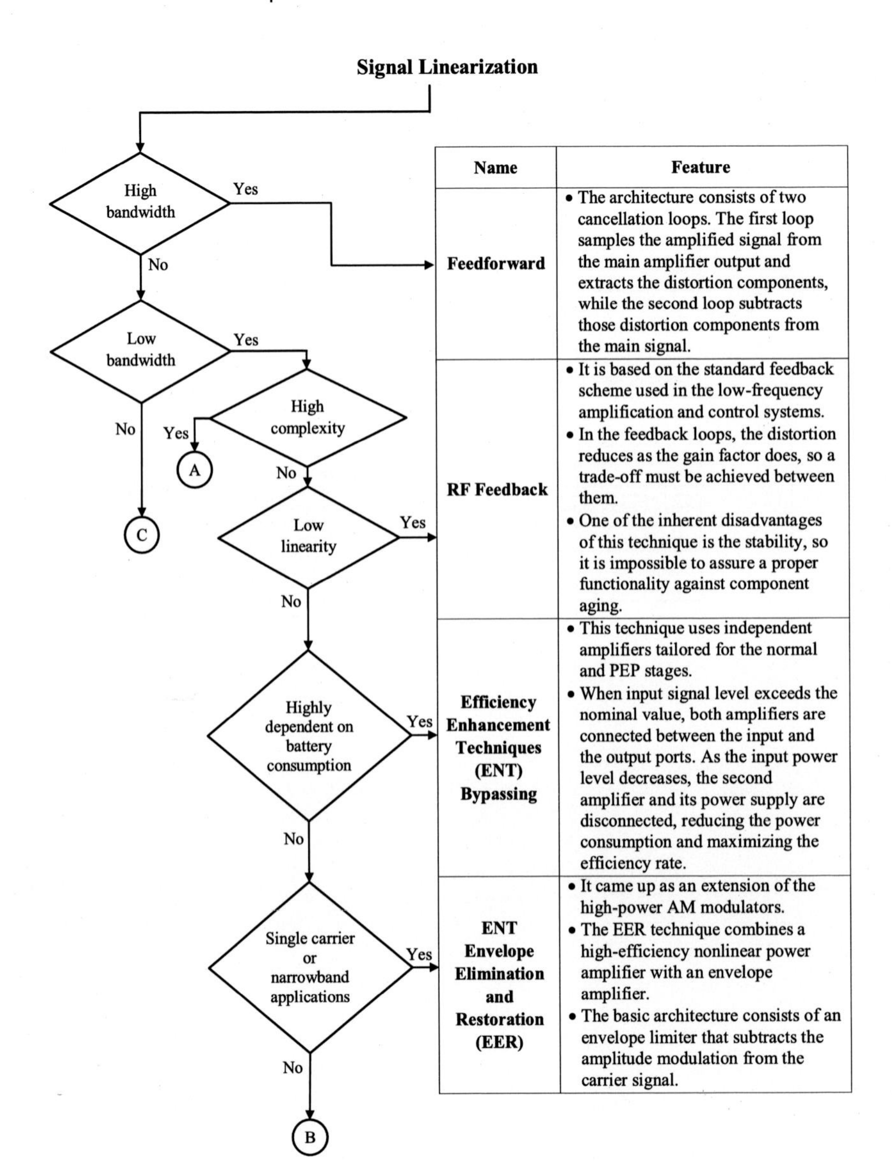

Name	Feature
Feedforward	• The architecture consists of two cancellation loops. The first loop samples the amplified signal from the main amplifier output and extracts the distortion components, while the second loop subtracts those distortion components from the main signal.
RF Feedback	• It is based on the standard feedback scheme used in the low-frequency amplification and control systems. • In the feedback loops, the distortion reduces as the gain factor does, so a trade-off must be achieved between them. • One of the inherent disadvantages of this technique is the stability, so it is impossible to assure a proper functionality against component aging.
Efficiency Enhancement Techniques (ENT) Bypassing	• This technique uses independent amplifiers tailored for the normal and PEP stages. • When input signal level exceeds the nominal value, both amplifiers are connected between the input and the output ports. As the input power level decreases, the second amplifier and its power supply are disconnected, reducing the power consumption and maximizing the efficiency rate.
ENT Envelope Elimination and Restoration (EER)	• It came up as an extension of the high-power AM modulators. • The EER technique combines a high-efficiency nonlinear power amplifier with an envelope amplifier. • The basic architecture consists of an envelope limiter that subtracts the amplitude modulation from the carrier signal.

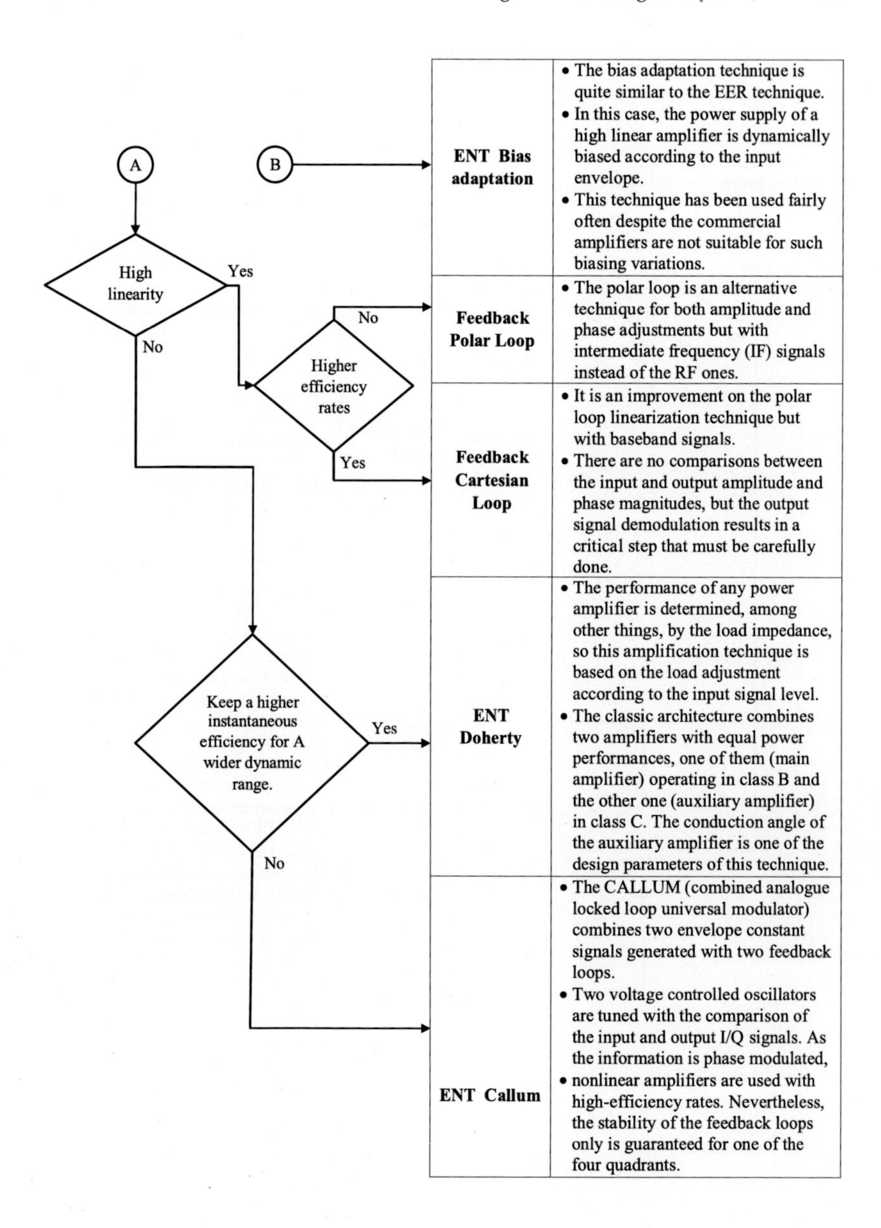

A
B
High linearity
Yes
No
Higher efficiency rates
No
Yes
Keep a higher instantaneous efficiency for A wider dynamic range.
Yes
No
ENT Bias adaptation
• The bias adaptation technique is quite similar to the EER technique.
• In this case, the power supply of a high linear amplifier is dynamically biased according to the input envelope.
• This technique has been used fairly often despite the commercial amplifiers are not suitable for such biasing variations.
Feedback Polar Loop
• The polar loop is an alternative technique for both amplitude and phase adjustments but with intermediate frequency (IF) signals instead of the RF ones.
Feedback Cartesian Loop
• It is an improvement on the polar loop linearization technique but with baseband signals.
• There are no comparisons between the input and output amplitude and phase magnitudes, but the output signal demodulation results in a critical step that must be carefully done.
ENT Doherty
• The performance of any power amplifier is determined, among other things, by the load impedance, so this amplification technique is based on the load adjustment according to the input signal level.
• The classic architecture combines two amplifiers with equal power performances, one of them (main amplifier) operating in class B and the other one (auxiliary amplifier) in class C. The conduction angle of the auxiliary amplifier is one of the design parameters of this technique.
ENT Callum
• The CALLUM (combined analogue locked loop universal modulator) combines two envelope constant signals generated with two feedback loops.
• Two voltage controlled oscillators are tuned with the comparison of the input and output I/Q signals. As the information is phase modulated,
• nonlinear amplifiers are used with high-efficiency rates. Nevertheless, the stability of the feedback loops only is guaranteed for one of the four quadrants.

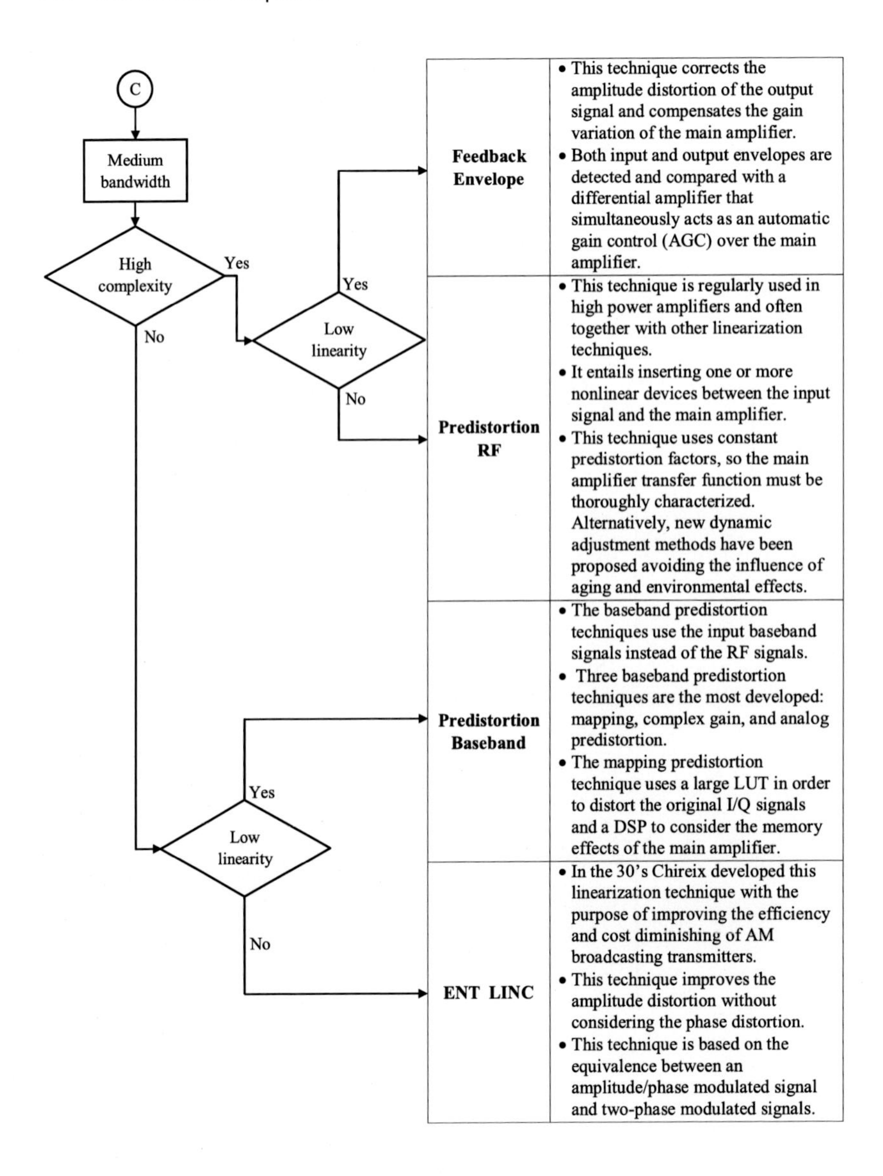
C
Medium bandwidth
High complexity
Yes
No
Low linearity
Yes
No
Low linearity
Yes
No
Feedback Envelope
• This technique corrects the amplitude distortion of the output signal and compensates the gain variation of the main amplifier.
• Both input and output envelopes are detected and compared with a differential amplifier that simultaneously acts as an automatic gain control (AGC) over the main amplifier.
Predistortion RF
• This technique is regularly used in high power amplifiers and often together with other linearization techniques.
• It entails inserting one or more nonlinear devices between the input signal and the main amplifier.
• This technique uses constant predistortion factors, so the main amplifier transfer function must be thoroughly characterized. Alternatively, new dynamic adjustment methods have been proposed avoiding the influence of aging and environmental effects.
Predistortion Baseband
• The baseband predistortion techniques use the input baseband signals instead of the RF signals.
• Three baseband predistortion techniques are the most developed: mapping, complex gain, and analog predistortion.
• The mapping predistortion technique uses a large LUT in order to distort the original I/Q signals and a DSP to consider the memory effects of the main amplifier.
ENT LINC
• In the 30's Chireix developed this linearization technique with the purpose of improving the efficiency and cost diminishing of AM broadcasting transmitters.
• This technique improves the amplitude distortion without considering the phase distortion.
• This technique is based on the equivalence between an amplitude/phase modulated signal and two-phase modulated signals.

Fourier Transform

From signal processing to function determination approaches, Fourier series are used in a broad range. The ease in operation is dependent on whether the function of odd or even.

Step 1: Fourier Series Approach of the Function		
Input	**Formula**	**Output**
f(x)= Given input function	$f(x)=a_0+\sum_{n=1}^{\infty}(an * \cos\left(\frac{n*\pi*x}{L}\right) + bn * \sin\left(\frac{n*\pi*x}{L}\right))$	f(x) as a periodic function

Step 2: Determining the Type of the Function		
Input	**Formula**	**Output**
f(x) as in Fourier series	$a_0=a_n=0$, if f(x) is odd $b_n=0$, if f(x) is even	a_0, a_n, and b_n

Step 3: Finding a_0		
Input	**Formula**	**Output**
f(x)=Function's itself L=Interval of periodicity	$a_0=\frac{1}{2L}\int_{-L}^{L} f(x)dx$	a_0=the first element of the Fourier series

Step 4: Finding an and bn		
Input	**Formula**	**Output**
f(x)=Function's itself L=Interval of periodicity	$a_n=\frac{1}{L}\int_{-L}^{L} f(x) * \cos\left(\frac{n*\pi*x}{L}\right)dx$ $b_n=\frac{1}{L}\int_{-L}^{L} f(x) * \sin\left(\frac{n*\pi*x}{L}\right)dx$	a_n=coefficient of cosine term b_n=coefficient of sine term

Step 5: Applying the Solution Set for an and bn		
Input	**Formula**	**Output**
$a_n=\frac{1}{L}\int_{-L}^{L} f(x) * \cos\left(\frac{n*\pi*x}{L}\right)dx$ $b_n=\frac{1}{L}\int_{-L}^{L} f(x) * \sin\left(\frac{n*\pi*x}{L}\right)dx$	By applying the integral solution for a_n and b_n, then we have a series approximation for the given function.	a_n and b_n

Step 6: Gathering the Obtained Parameters		
Input	**Formula**	**Output**
a_0, a_n, b_n and n value	$f(x)=a_0+\sum_{n=1}^{\infty}(an * \cos\left(\frac{n*\pi*x}{L}\right) + bn * \sin\left(\frac{n*\pi*x}{L}\right))$	f(x) as Fourier series approach

Amplifiers

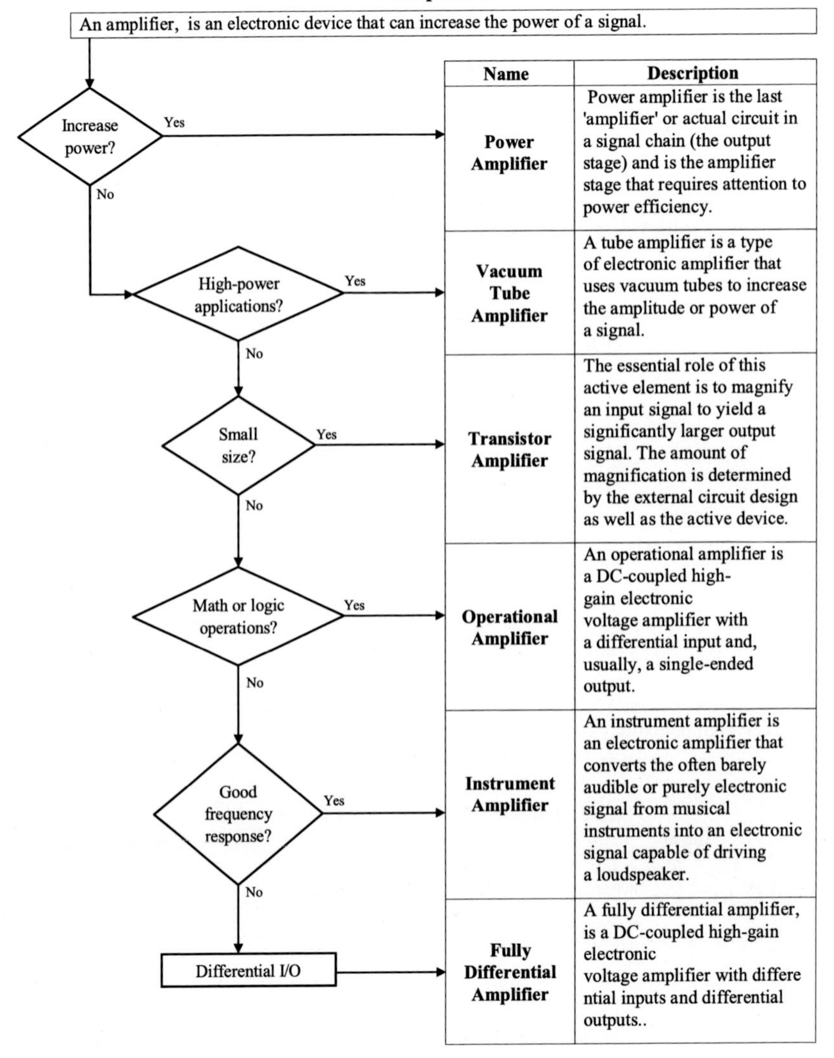

Name	Description
Power Amplifier	Power amplifier is the last 'amplifier' or actual circuit in a signal chain (the output stage) and is the amplifier stage that requires attention to power efficiency.
Vacuum Tube Amplifier	A tube amplifier is a type of electronic amplifier that uses vacuum tubes to increase the amplitude or power of a signal.
Transistor Amplifier	The essential role of this active element is to magnify an input signal to yield a significantly larger output signal. The amount of magnification is determined by the external circuit design as well as the active device.
Operational Amplifier	An operational amplifier is a DC-coupled high-gain electronic voltage amplifier with a differential input and, usually, a single-ended output.
Instrument Amplifier	An instrument amplifier is an electronic amplifier that converts the often barely audible or purely electronic signal from musical instruments into an electronic signal capable of driving a loudspeaker.
Fully Differential Amplifier	A fully differential amplifier, is a DC-coupled high-gain electronic voltage amplifier with differe ntial inputs and differential outputs..

Calculation for Sampling Rate

In signal processing, sampling is the reduction of a continuous signal to a discrete signal. A common example is the conversion of a sound wave (a continuous signal) to a sequence of samples (a discrete-time signal).

Step 1

Input	Formula	Output
Max frequency content of the signal	**Nyquist Sampling Theorem** Select a sampling frequency twice as estimated max frequency content of the signal	Sampling frequency

Step 2

Input	Formula	Output
Input in general: time-domain signal samples N: Number of samples (must be a power of 2) x: Input in time domain m: Time-domain sample index	**Fast Fourier Transform** $X_k = \sum_{m=0}^{\frac{N}{2}-1} x_{2m} e^{-\frac{4\pi i}{N}mk} + e^{-\frac{2\pi i}{N}k} \sum_{m=0}^{\frac{N}{2}-1} x_{2m+1} e^{-\frac{4\pi i}{N}mk}$ $X_{k+\frac{N}{2}} = \sum_{m=0}^{\frac{N}{2}-1} x_{2m} e^{-\frac{4\pi i}{N}mk} - e^{-\frac{2\pi i}{N}k} \sum_{m=0}^{\frac{N}{2}-1} x_{2m+1} e^{-\frac{4\pi i}{N}mk}$	Output in general: frequency-domain spectrum of the input signal X: Output in frequency domain k: Frequency-domain element index $(0 \le k < \frac{N}{2})$

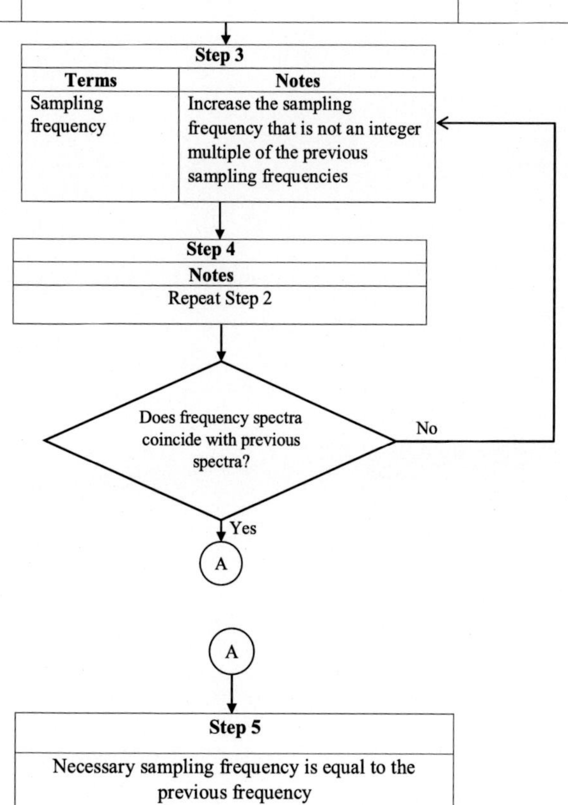

Chapter 14

Controls Theory and Applications

ABSTRACT

Any mechanical system consists of three basic parts: mechanical, electronics, and software. All of these three systems have tolerances. Mechanical parts have both dimensional and shape tolerances, because it would be too costly to manufacture mechanical parts without any tolerances. The tolerances show the acceptable error in the part. Because of the tolerances in dimension, the mechanical model of the system cannot be exact; if the part is bigger, it will be heavier in the tolerance limits. Electrical parts also have tolerances; for example, a resistor cannot have an exact value—it is also manufactured with tolerances. This means the signal going over this resistance will also have a variation. In software, there are also tolerances. For example, one a number is defined as floating or long so it is represented slightly different. There is also truncation error in the software. When all these three kinds of tolerances and errors add up with the noise in the system, it is not possible for the system to act as in theory. This is why mechatronic systems need control. The control of the system is achieved with software mostly inside a microcontroller. The system receives feedback signals from the sensors and with the predefined calculations responds with the actuators and tries to achieve the desired purpose.

Mechatronic Components. https://doi.org/10.1016/B978-0-12-814126-7.00014-1

Laplace Transform

Laplace Transform is applied for time-domain systems, such as automated control systems and partial differential equations.

Step 1: System Mathematical Modeling		
Input	Formula	Output
Any ordinary differential equation of elements: $y=f(t), \frac{dy}{dt}, \frac{d^2y}{dt^2}, \frac{d^ny}{dt^n}$, r=f(t)	$a_n\frac{d^ny}{dt^n} + a_{n-1}\frac{d^{n-1}y}{dt^{n-1}} + ...a1y = bn\frac{d^nr}{dt^n} + ...$	Obtained mathematical model

Step 2: Turning Time-Domain into s-Domain		
Input	Formula	Output
Obtained mathematical model	L{y(t)}=Y(s) For more, see Laplace Table	$(a_ns^n + a_{n-1}s^{n-1} + ...a_1s)Y(s) = (b_ns^n + ...)R(s)$

Step 3: Obtaining Transfer Function		
Input	Formula	Output
Related DE of s-domain	$\frac{Y(s)}{R(s)} = \frac{(bnsn + ...)}{(ansn + an-1sn-1 + ...a1s)} = H(s)$	H(s)=Transfer function

Step 4: Applying the Input Function and Leaving Y(s) Alone		
Input	Formula	Output
H(s)=Transfer function R(s)=Input function	$Y(s) = \frac{(bnsn + ...)}{(ansn + an-1sn-1 + ...a1s)}R(s)$	Y(s)=Output function

Step 5: Applying Inverse Laplace Transform to Find y(t)		
Input	Formula	Output
Y(s)=Output function	$L^{-1}\{Y(s)\}=y(t)$	y(t)=Output function in time-domain

Conversion From State Space To Transfer Function

Conversion from state space to transfer function will be done according to the system linearity.

Step 1	
To Do	**Output**
Define the mathematical model of the system	• Differential equations of the system

Step 2	
Input	**Output**
Write down the differential equation of the system and assign the state variables	• A= State matrix • B=Input matrix • C=Output matrix • D=Feedforward matrix • [X]=State variable matrix • U=Applied input (force, torque, voltage, etc.)

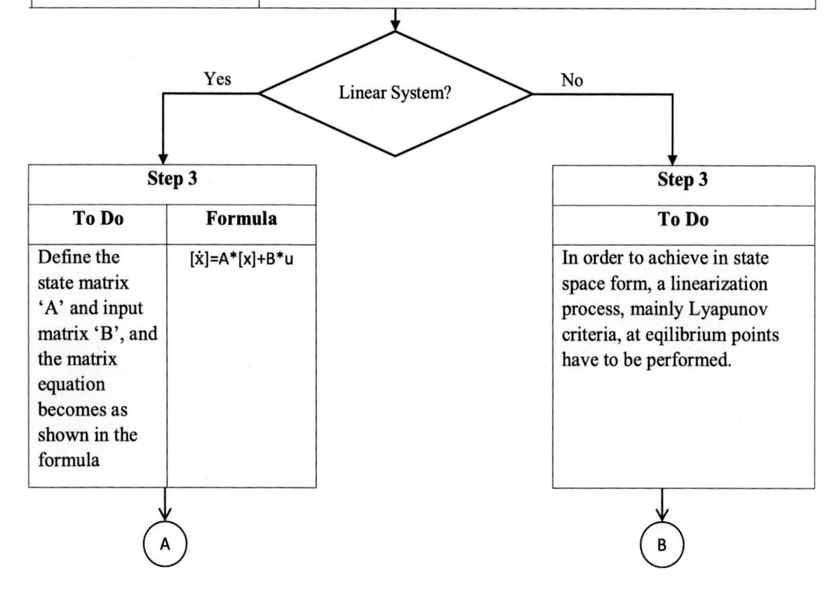

A

Step 4	
To Do	**Formula**
Define the matrices 'C' and 'D', so the second matrix equation becomes as shown in the formula	$[y]=C*[x]+D*u$

Step 5	
To Do	**Formula**
Calculate the respective matrix and determinant operation such as shown in the formula	$H(s)=\|[C*(sI-A)^{-1}*B]\|$

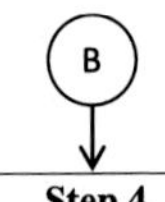

Step 4	
To Do	**Formula**
Linearize the system at equilibrium points, and perform 'A' and 'B' matrices, such as shown in the formula	$[\Delta\dot{x}]=A*[\Delta x]+B*u$

Step 5	
To Do	**Formula**
If input is also nonlinear, linearize again at equilibrium points, and perform 'C' and 'D' matrices, such as shown in the formula	$[\Delta y]=C*[\Delta x]+D*u$

Step 6	
To Do	**Formula**
Calculate the respective matrix and determinant operation such as shown in the formula	$H(s)=\|[C*(sI-A)^{-1}*B]\|$

Conversion From Transfer Function To State Space

We here consider a system defined by

$$y^{(n)} + a_1 y^{(n-1)} + \cdots + a_{n-1}\dot{y} + a_n y = b_0 u^{(n)} + b_1 u^{(n-1)} + \cdots + b_{n-1}\dot{u} + b_n u\,, \tag{1}$$

where u is the control input and y is the output. We can write this equation as

$$\frac{Y(s)}{U(s)} = \frac{b_0 s^n + b_1 s^{n-1} + \cdots + b_{n-1}s + b_n}{s^n + a_1 s^{n-1} + \cdots + a_{n-1}s + a_n}\,. \tag{2}$$

Later, we shall present state-space representation of the system defined by (1) and (2) in controllable canonical form (A), observable canonical form (B), and diagonal canonical form (C).

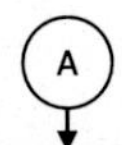

Controllable Canonical Form	
Note	Formula
Note that the controllable canonical form is important in discussing the pole-placement approach to the control system design.	$\begin{bmatrix} \dot{x}_1 \\ \dot{x}_2 \\ \vdots \\ \dot{x}_{n-1} \\ \dot{x}_n \end{bmatrix} = \begin{bmatrix} 0 & 1 & 0 & \cdots & 0 \\ 0 & 0 & 1 & \cdots & 0 \\ \vdots & \vdots & \vdots & \ddots & \vdots \\ 0 & 0 & 0 & \cdots & 1 \\ -a_n & -a_{n-1} & -a_{n-2} & \cdots & -a_1 \end{bmatrix} \begin{bmatrix} x_1 \\ x_2 \\ \vdots \\ x_{n-1} \\ x_n \end{bmatrix} + \begin{bmatrix} 0 \\ 0 \\ \vdots \\ 0 \\ 1 \end{bmatrix} u \quad (3)$ $y = [(b_n - a_n b_0) \quad (b_{n-1} - a_{n-1} b_0) \quad \ldots \quad (b_1 - a_1 b_0)] \begin{bmatrix} x_1 \\ x_2 \\ \vdots \\ x_{n-1} \\ x_n \end{bmatrix} + b_0 u \quad (4)$

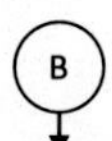

Observable Canonical Form	
Note	Formula
We consider the following state-space representation, being called an observable canonical form, as shown in the formula.	$\begin{bmatrix} \dot{x}_1 \\ \dot{x}_2 \\ \vdots \\ \dot{x}_{n-1} \\ \dot{x}_n \end{bmatrix} = \begin{bmatrix} 0 & 1 & 0 & \cdots & 0 \\ 0 & 0 & 1 & \cdots & 0 \\ \vdots & \vdots & \vdots & \ddots & \vdots \\ 0 & 0 & 0 & \cdots & 1 \\ -a_n & -a_{n-1} & -a_{n-2} & \cdots & -a_1 \end{bmatrix}^T \begin{bmatrix} x_1 \\ x_2 \\ \vdots \\ x_{n-1} \\ x_n \end{bmatrix} + \begin{bmatrix} b_n - a_n b_0 \\ b_{n-1} - a_{n-1} b_0 \\ \vdots \\ \vdots \\ b_1 - a_1 b_0 \end{bmatrix} u$ $y = [0 \quad 0 \quad 0 \quad \ldots \quad 1] \begin{bmatrix} x_1 \\ x_2 \\ \vdots \\ x_{n-1} \\ x_n \end{bmatrix} + b_0 u$

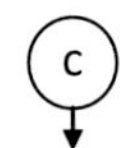

Diagonal Canonical Form	
Step 1	
Note	**Formula**
We here consider the transfer function system given by (2). We have the case where the dominator polynomial involves only distinct roots. For the distinct root case, we can write (2) in the form shown in (3) and (4)	$\frac{Y(s)}{U(s)} = \frac{b_0 s^n + b_1 s^{n-1} + \cdots + b_{n-1}s + b_n}{(s+p_1)(s+p_2)\ldots(s+p_n)}$ (3) $= b_0 + \frac{c_1}{s+p_1} + \frac{c_2}{s+p_2} + \cdots + \frac{c_n}{s+p_n}$ (4)

Step 2	
Note	**Formula**
The diagonal canonical form of the state-space representation of this system is given by the formula	$\begin{bmatrix} \dot{x}_1 \\ \dot{x}_2 \\ \vdots \\ \dot{x}_{n-1} \\ \dot{x}_n \end{bmatrix} = \begin{bmatrix} -p_n & 0 & \cdots & 0 & 0 \\ 0 & -p_n & \cdots & 0 & 0 \\ \vdots & \vdots & \ddots & \vdots & \vdots \\ 0 & 0 & \cdots & -p_n & 0 \\ 0 & 0 & \cdots & 0 & -p_n \end{bmatrix} \begin{bmatrix} x_1 \\ x_2 \\ \vdots \\ x_{n-1} \\ x_n \end{bmatrix} + \begin{bmatrix} 1 \\ 1 \\ \vdots \\ 1 \\ 1 \end{bmatrix} u$ $y = [c_1 \quad c_2 \quad c_3 \quad \cdots \quad c_n] \begin{bmatrix} x_1 \\ x_2 \\ \vdots \\ x_{n-1} \\ x_n \end{bmatrix} + b_0 u$

Control Structures

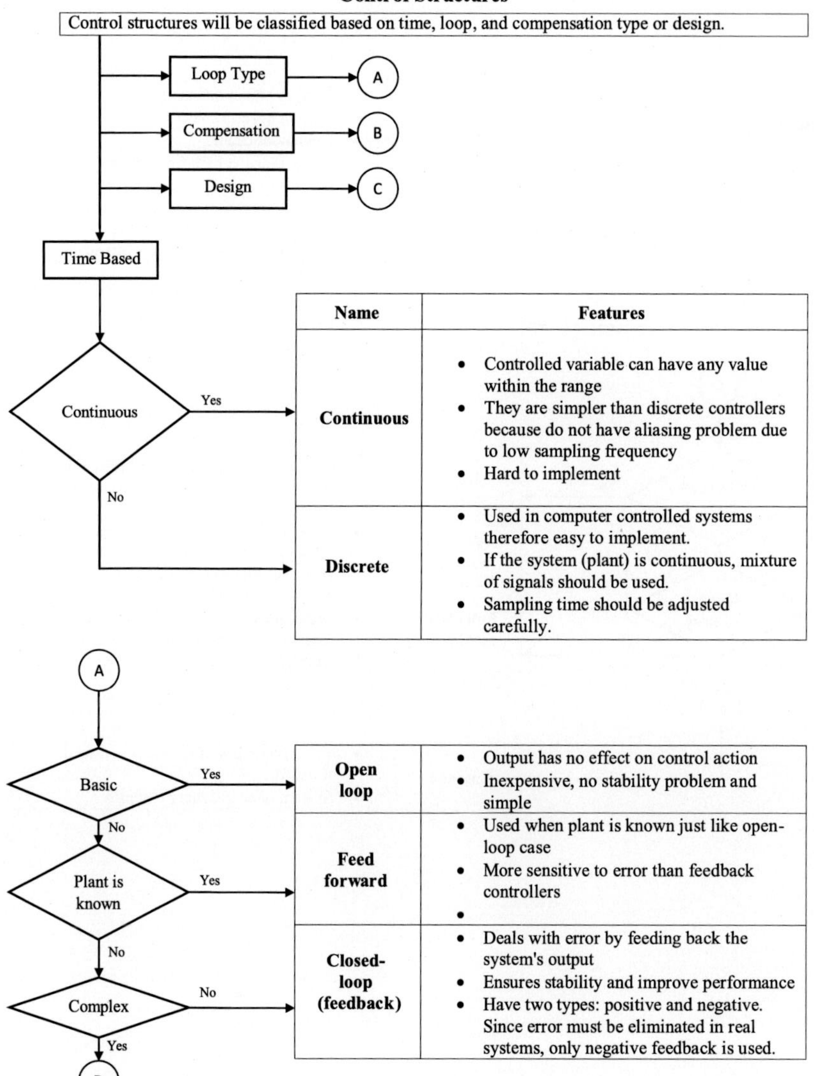

Name	Features
Continuous	• Controlled variable can have any value within the range • They are simpler than discrete controllers because do not have aliasing problem due to low sampling frequency • Hard to implement
Discrete	• Used in computer controlled systems therefore easy to implement. • If the system (plant) is continuous, mixture of signals should be used. • Sampling time should be adjusted carefully.

Open loop	• Output has no effect on control action • Inexpensive, no stability problem and simple
Feed forward	• Used when plant is known just like open-loop case • More sensitive to error than feedback controllers •
Closed-loop (feedback)	• Deals with error by feeding back the system's output • Ensures stability and improve performance • Have two types: positive and negative. Since error must be eliminated in real systems, only negative feedback is used.

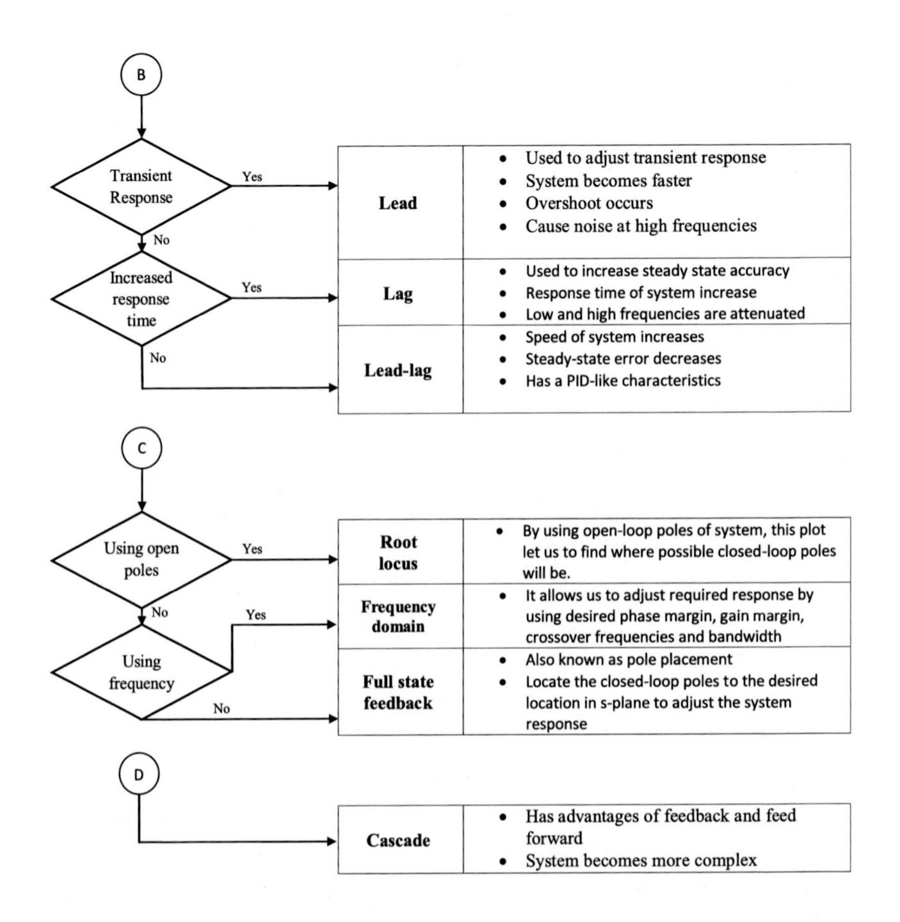
B
Transient Response
Yes
No
Increased response time
Yes
No
Lead
Used to adjust transient response
System becomes faster
Overshoot occurs
Cause noise at high frequencies
Lag
Used to increase steady state accuracy
Response time of system increase
Low and high frequencies are attenuated
Lead-lag
Speed of system increases
Steady-state error decreases
Has a PID-like characteristics
C
Using open poles
Yes
No
Using frequency
Yes
No
Root locus
By using open-loop poles of system, this plot let us to find where possible closed-loop poles will be.
Frequency domain
It allows us to adjust required response by using desired phase margin, gain margin, crossover frequencies and bandwidth
Full state feedback
Also known as pole placement
Locate the closed-loop poles to the desired location in s-plane to adjust the system response
D
Cascade
Has advantages of feedback and feed forward
System becomes more complex

Basic Feedback Controller Types

A feedback controller is a device or algorithm running on a microcontroller, which regularly examines the process it is in charge of, in order to make changes that will improve its output efficiency.

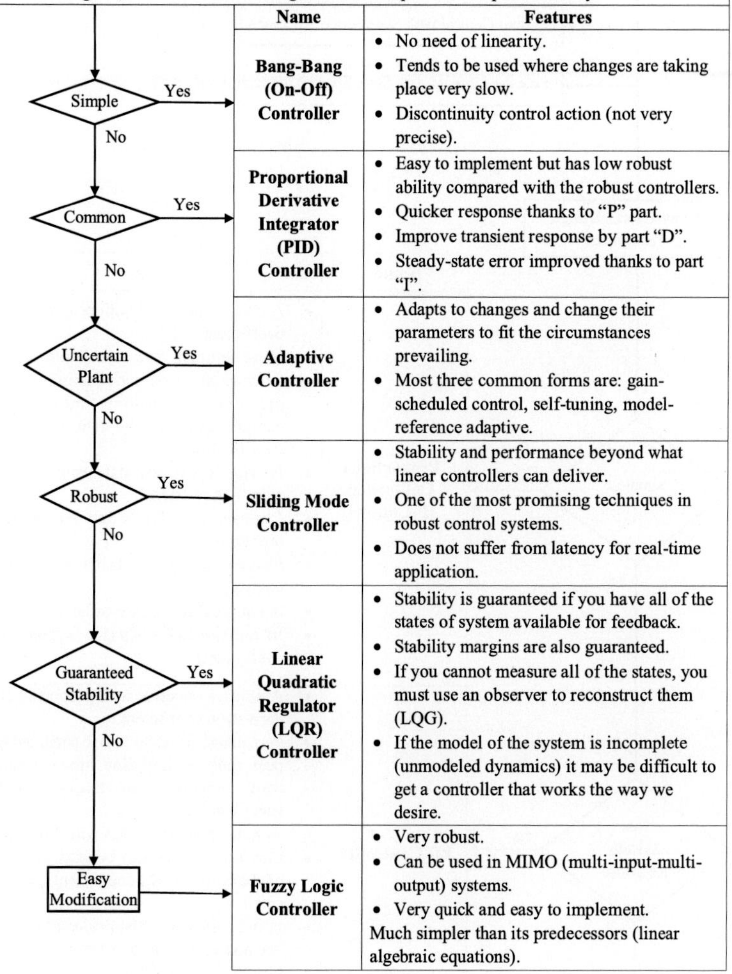

Name	Features
Bang-Bang (On-Off) Controller	• No need of linearity. • Tends to be used where changes are taking place very slow. • Discontinuity control action (not very precise).
Proportional Derivative Integrator (PID) Controller	• Easy to implement but has low robust ability compared with the robust controllers. • Quicker response thanks to "P" part. • Improve transient response by part "D". • Steady-state error improved thanks to part "I".
Adaptive Controller	• Adapts to changes and change their parameters to fit the circumstances prevailing. • Most three common forms are: gain-scheduled control, self-tuning, model-reference adaptive.
Sliding Mode Controller	• Stability and performance beyond what linear controllers can deliver. • One of the most promising techniques in robust control systems. • Does not suffer from latency for real-time application.
Linear Quadratic Regulator (LQR) Controller	• Stability is guaranteed if you have all of the states of system available for feedback. • Stability margins are also guaranteed. • If you cannot measure all of the states, you must use an observer to reconstruct them (LQG). • If the model of the system is incomplete (unmodeled dynamics) it may be difficult to get a controller that works the way we desire.
Fuzzy Logic Controller	• Very robust. • Can be used in MIMO (multi-input-multi-output) systems. • Very quick and easy to implement. Much simpler than its predecessors (linear algebraic equations).

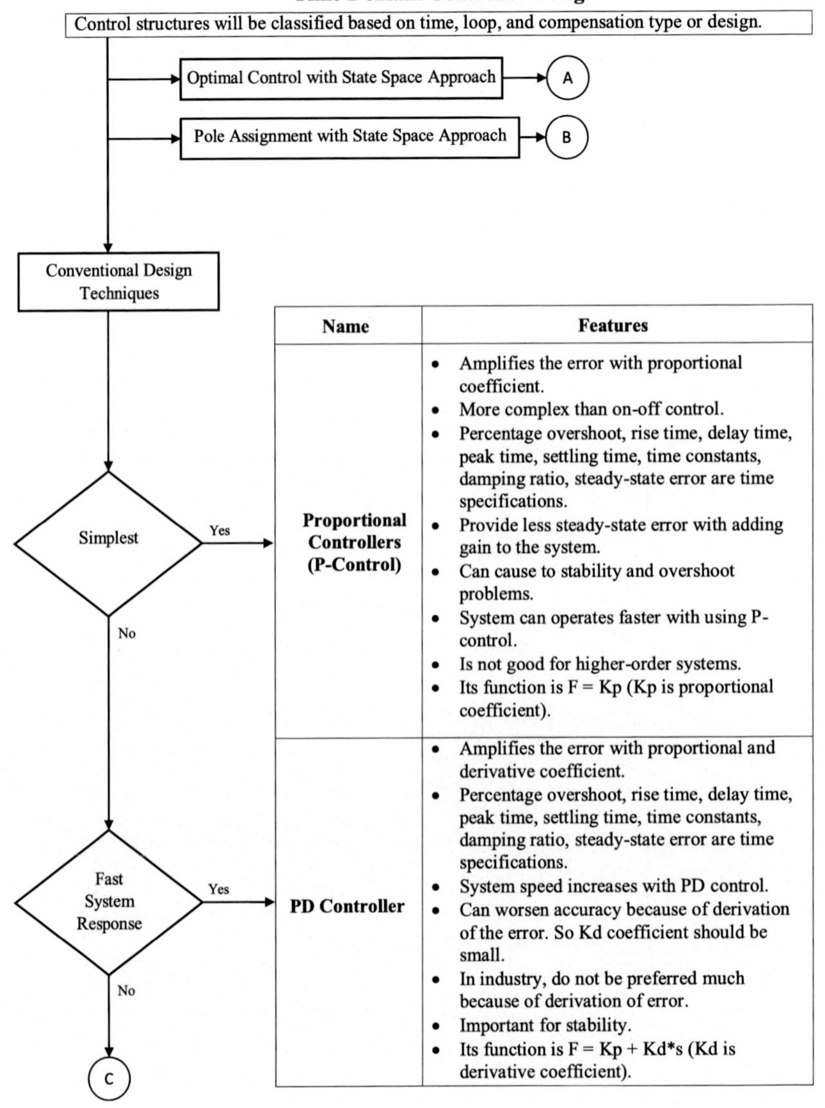

Name	Features
Proportional Controllers (P-Control)	• Amplifies the error with proportional coefficient. • More complex than on-off control. • Percentage overshoot, rise time, delay time, peak time, settling time, time constants, damping ratio, steady-state error are time specifications. • Provide less steady-state error with adding gain to the system. • Can cause to stability and overshoot problems. • System can operates faster with using P-control. • Is not good for higher-order systems. • Its function is F = Kp (Kp is proportional coefficient).
PD Controller	• Amplifies the error with proportional and derivative coefficient. • Percentage overshoot, rise time, delay time, peak time, settling time, time constants, damping ratio, steady-state error are time specifications. • System speed increases with PD control. • Can worsen accuracy because of derivation of the error. So Kd coefficient should be small. • In industry, do not be preferred much because of derivation of error. • Important for stability. • Its function is F = Kp + Kd*s (Kd is derivative coefficient).

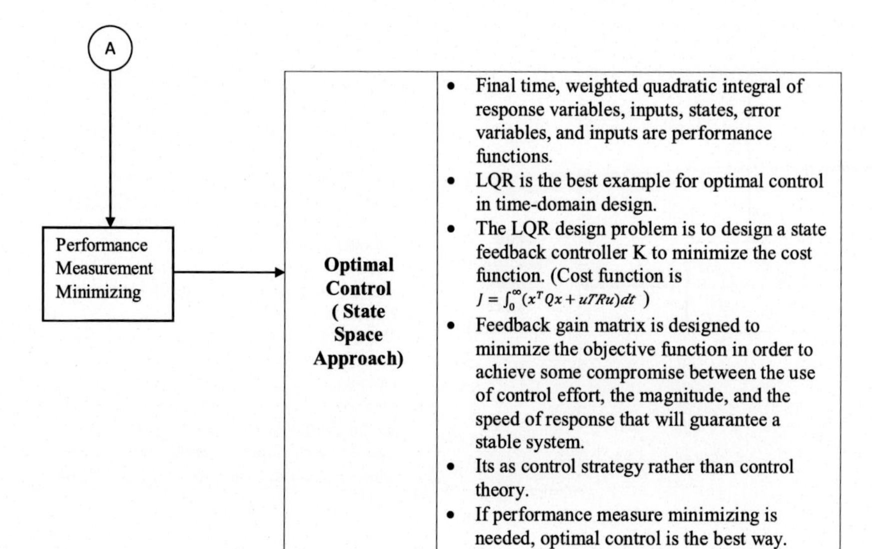
A
Performance
Measurement
Minimizing
Optimal
Control
(State
Space
Approach)
Final time, weighted quadratic integral of response variables, inputs, states, error variables, and inputs are performance functions.
LQR is the best example for optimal control in time-domain design.
The LQR design problem is to design a state feedback controller K to minimize the cost function. (Cost function is $J = \int_0^\infty (x^T Qx + uTRu)dt$)
Feedback gain matrix is designed to minimize the objective function in order to achieve some compromise between the use of control effort, the magnitude, and the speed of response that will guarantee a stable system.
Its as control strategy rather than control theory.
If performance measure minimizing is needed, optimal control is the best way.

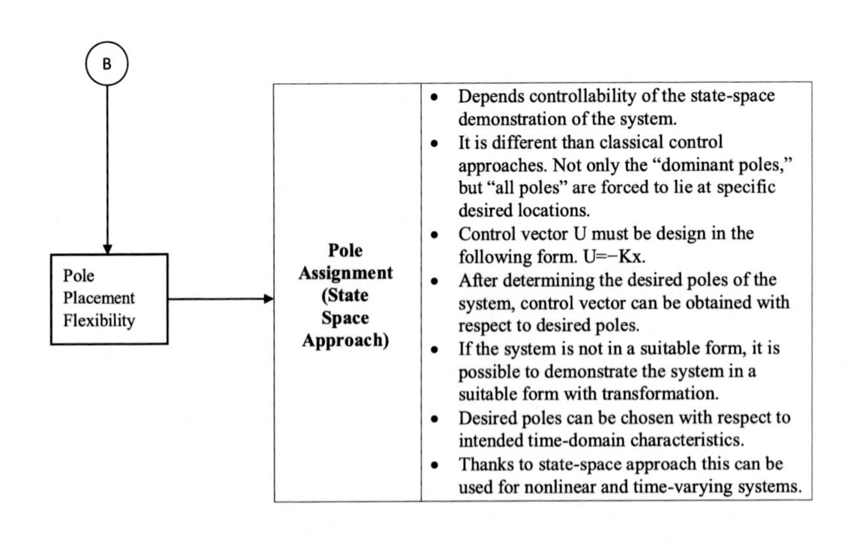
B
Pole
Placement
Flexibility
Pole
Assignment
(State
Space
Approach)
Depends controllability of the state-space demonstration of the system.
It is different than classical control approaches. Not only the "dominant poles," but "all poles" are forced to lie at specific desired locations.
Control vector U must be design in the following form. U=−Kx.
After determining the desired poles of the system, control vector can be obtained with respect to desired poles.
If the system is not in a suitable form, it is possible to demonstrate the system in a suitable form with transformation.
Desired poles can be chosen with respect to intended time-domain characteristics.
Thanks to state-space approach this can be used for nonlinear and time-varying systems.

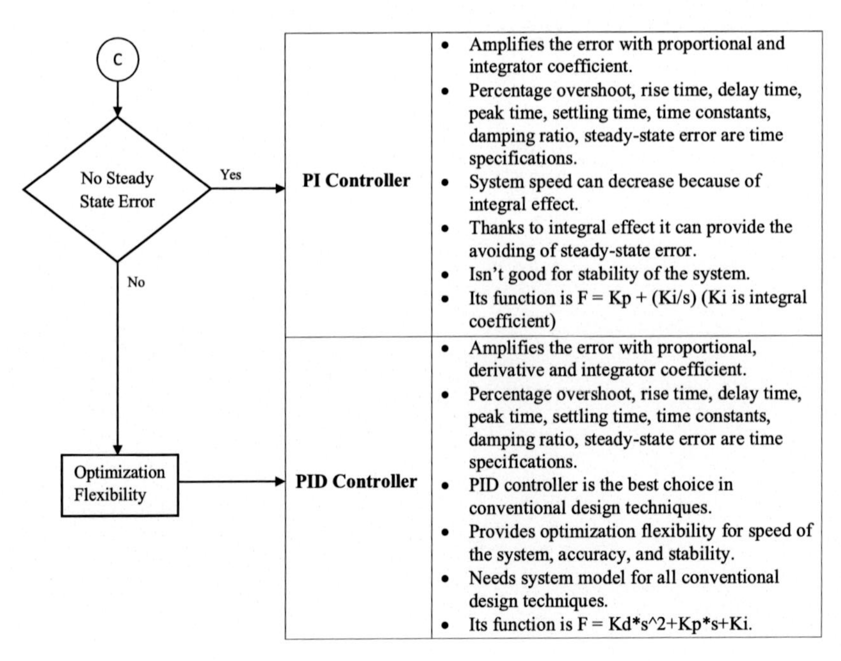
C
No Steady
State Error
Yes
No
PI Controller
Amplifies the error with proportional and integrator coefficient.
Percentage overshoot, rise time, delay time, peak time, settling time, time constants, damping ratio, steady-state error are time specifications.
System speed can decrease because of integral effect.
Thanks to integral effect it can provide the avoiding of steady-state error.
Isn't good for stability of the system.
Its function is F = Kp + (Ki/s) (Ki is integral coefficient)
Optimization
Flexibility
PID Controller
Amplifies the error with proportional, derivative and integrator coefficient.
Percentage overshoot, rise time, delay time, peak time, settling time, time constants, damping ratio, steady-state error are time specifications.
PID controller is the best choice in conventional design techniques.
Provides optimization flexibility for speed of the system, accuracy, and stability.
Needs system model for all conventional design techniques.
Its function is F = Kd*s^2+Kp*s+Ki.

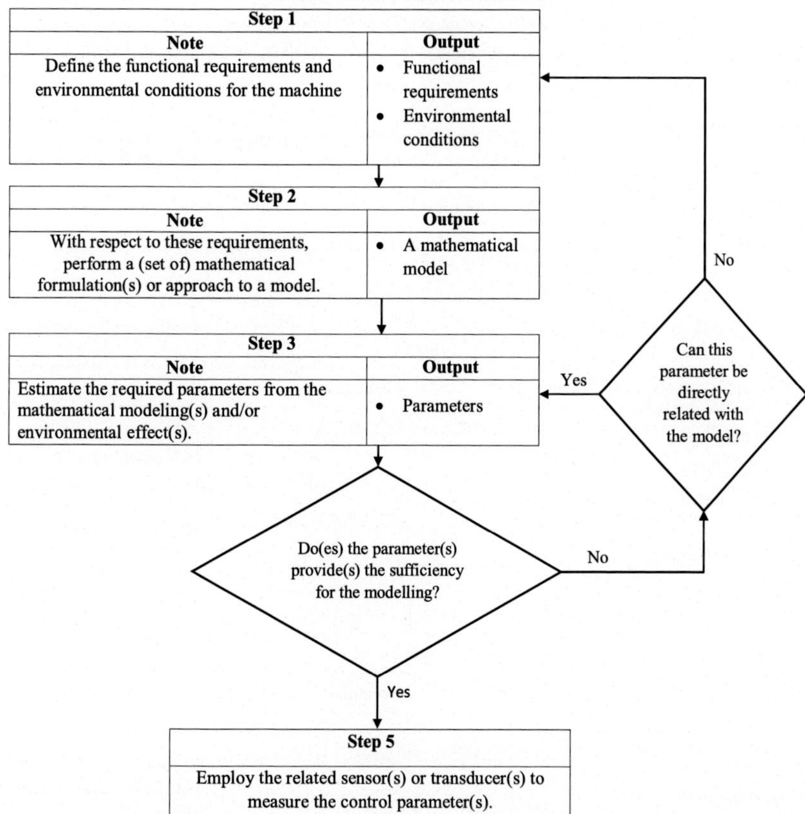
Selection of Control Parameters
Step 1
Note
Output
Define the functional requirements and environmental conditions for the machine
• Functional requirements
• Environmental conditions
Step 2
Note
Output
With respect to these requirements, perform a (set of) mathematical formulation(s) or approach to a model.
• A mathematical model
Step 3
Note
Output
Estimate the required parameters from the mathematical modeling(s) and/or environmental effect(s).
• Parameters
Yes
No
Can this parameter be directly related with the model?
Do(es) the parameter(s) provide(s) the sufficiency for the modelling?
No
Yes
Step 5
Employ the related sensor(s) or transducer(s) to measure the control parameter(s).

Lead Compensator Design

The transfer function of a phase-lead compensator: $G_c(s) = \alpha\frac{s+z}{s+p}$ where $|p| > |z|$

Lead compensator aims to improve transient response (overshoot, rise time, settling time, etc.) of a dynamic system.

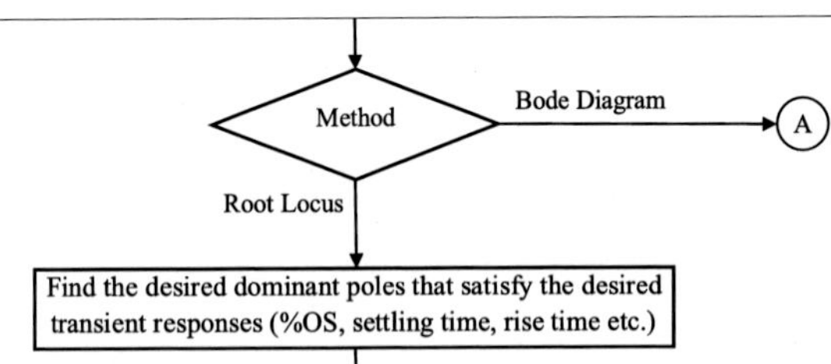

Step 1		
Input	**Formulas**	**Output**
T_s: Settling time T_p: Peak time T_r: Rise time %OS: Maximum overshoot percentage $\beta = \cos^{-1}\zeta$ ω_d: Damped frequency	$T_s = \frac{4}{\zeta\omega}$ $\quad T_p = \frac{\pi}{\omega_n\sqrt{1-\zeta^2}}$ $\zeta = -\frac{\ln\left(\%\frac{OS}{100}\right)}{\sqrt{\pi^2 + \ln^2\left(\%\frac{OS}{100}\right)}}$ $\omega_d = \omega_n\sqrt{1-\zeta^2}$ $\quad T_r = \frac{\pi-\beta}{\omega_d}$	ζ: Damping ratio ω_n: Natural frequency ω_d: Damped frequency

Step 2		
Input	**Formulas**	**Output**
ζ: Damping ratio ω_n: Natural frequency ω_d: Damped frequency	Imaginary ($j\omega$) Axis $s=\sigma+j\omega$ ω_n $\omega_d=\omega_n\sqrt{1-\zeta^2}$ $\cos^{-1}(\zeta)=\theta$ Real (σ) Axis $\zeta\omega_n$	σ : Horizontal axis component of the desired root on s-plane ω : Vertical axis component of the desired root on s-plane

B

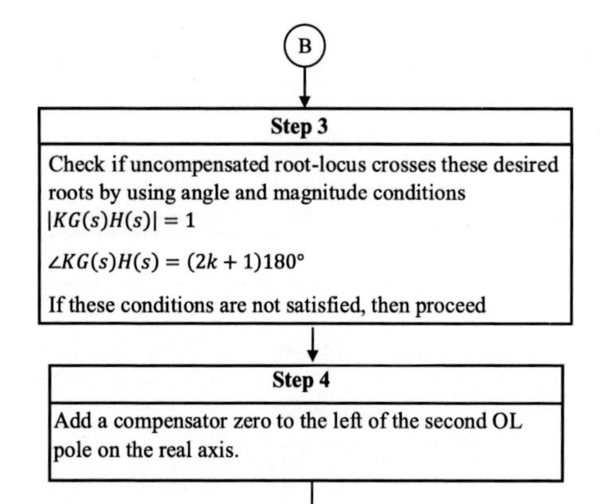

Step 5

Input	Formulas	Output
n: Number of OL poles m: Number of OL zeros θ_k: Angles of OL poles to desired pole ψ_p: Angles of OL zeros to desired pole ψ_c: Angle of added compensator zero to desired pole	$\sum_{k=1}^{n} \theta_k - \sum_{p=1}^{m} \psi_p = 180 + \psi_c - \theta_c$	θ_c: Angle of compensator pole to be added, to desired pole

Step 6

Find the compensator pole location on the real axis from trigonometry by using θ_c.

Step 7

Check the system response by simulating whether compensated system meets the system specisifications. If it does not, go to step 4

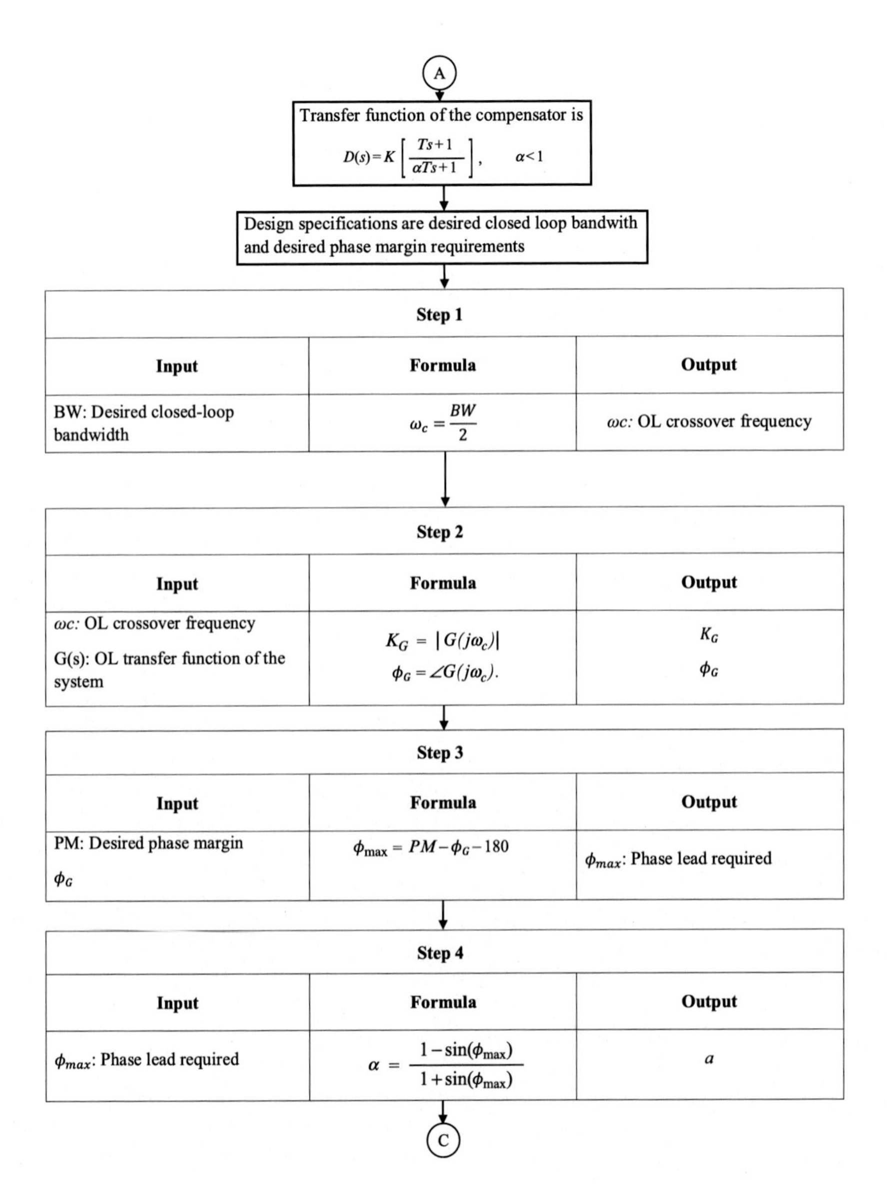
A
Transfer function of the compensator is
$D(s)=K\left[\frac{Ts+1}{\alpha Ts+1}\right], \quad \alpha<1$
Design specifications are desired closed loop bandwith and desired phase margin requirements
Step 1
Input
Formula
Output
BW: Desired closed-loop bandwidth
$\omega_c = \frac{BW}{2}$
ωc: OL crossover frequency
Step 2
Input
Formula
Output
ωc: OL crossover frequency
G(s): OL transfer function of the system
$K_G = |G(j\omega_c)|$
$\phi_G = \angle G(j\omega_c).$
K_G
ϕ_G
Step 3
Input
Formula
Output
PM: Desired phase margin
ϕ_G
$\phi_{\max} = PM - \phi_G - 180$
ϕ_{max}: Phase lead required
Step 4
Input
Formula
Output
ϕ_{max}: Phase lead required
$\alpha = \frac{1-\sin(\phi_{\max})}{1+\sin(\phi_{\max})}$
a
C

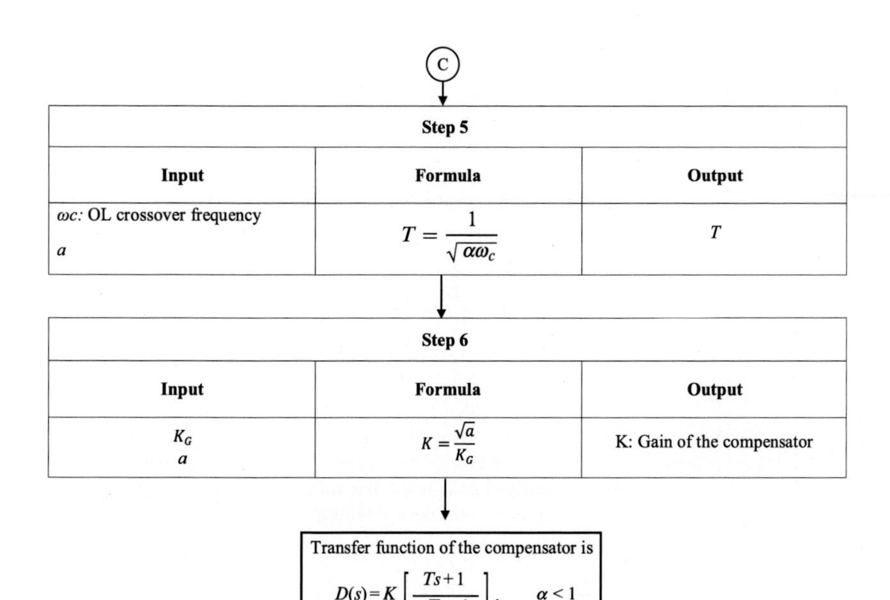

Step 5		
Input	**Formula**	**Output**
ωc: OL crossover frequency a	$T = \frac{1}{\sqrt{\alpha\omega_c}}$	T

Step 6		
Input	**Formula**	**Output**
K_G a	$K = \frac{\sqrt{a}}{K_G}$	K: Gain of the compensator

Transfer function of the compensator is

$$D(s) = K\left[\frac{Ts+1}{\alpha Ts+1}\right], \quad \alpha < 1$$

Lag Compensator Design

The transfer function of a phase-lag compensator: $G_c(s) = \alpha \frac{s+z}{s+p}$ where $|p| < |z|$
PI-type controller is a type of phase-lag compensator which has a pole at zero.

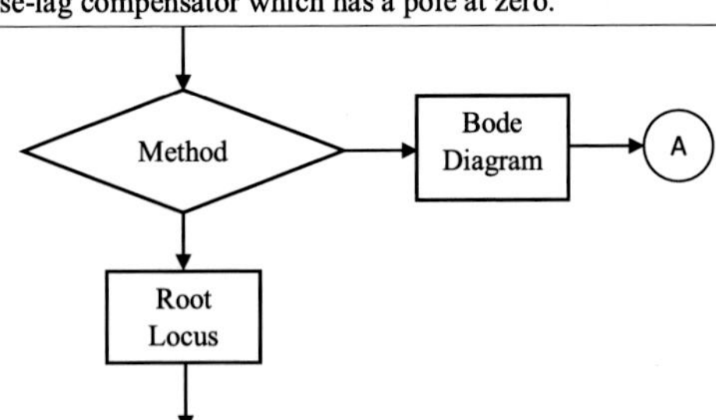

Step 1		
To Do	**How To?**	**Output**
Find the target regions	• Set ζ and w_n values for any complex poles to get a desirable percent overshoot and settling time	ζ: Damping ratio w_n: Natural frequency

Step 2		
To Do	**How To?**	**Output**
Examine the uncompensated root locus	• Examine the uncompensated RL diagram to see if the pole locations determined above can be met with only proportional control. • Find the desirable root locations and the associated gain K and determine if the error requirement is satisfied. If not, choose $\alpha = \frac{K_{uncomp}}{K_{desired}}$ and proceed.	K: Gain

Step 3	
To Do	**How To?**
Add a real zero and a real pole	Add a real zero and a real pole to GH(s) modifying the root locus diagram to move branches into the target region.

Step 4
To Do
Simulate the Time-domain performance

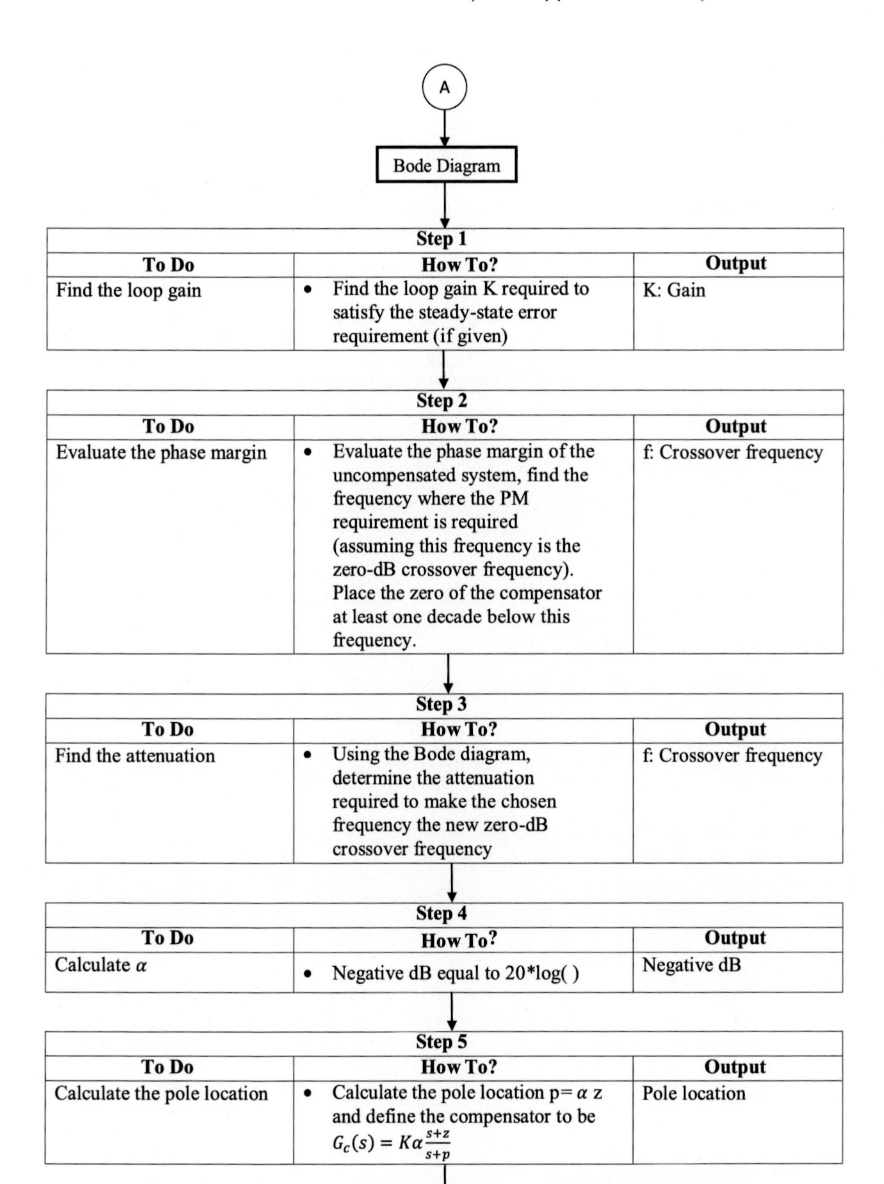

Step 1		
To Do	**How To?**	**Output**
Find the loop gain	• Find the loop gain K required to satisfy the steady-state error requirement (if given)	K: Gain

Step 2		
To Do	**How To?**	**Output**
Evaluate the phase margin	• Evaluate the phase margin of the uncompensated system, find the frequency where the PM requirement is required (assuming this frequency is the zero-dB crossover frequency). Place the zero of the compensator at least one decade below this frequency.	f: Crossover frequency

Step 3		
To Do	**How To?**	**Output**
Find the attenuation	• Using the Bode diagram, determine the attenuation required to make the chosen frequency the new zero-dB crossover frequency	f: Crossover frequency

Step 4		
To Do	**How To?**	**Output**
Calculate α	• Negative dB equal to 20*log( )	Negative dB

Step 5		
To Do	**How To?**	**Output**
Calculate the pole location	• Calculate the pole location p= α z and define the compensator to be $G_c(s) = K\alpha\frac{s+z}{s+p}$	Pole location

B

Step 6	
To Do	**How To?**
Check the phase margin	Check the PM of compensated system to see if desired value has been attained. If not, then decide on the additional phase required by the compensator.

Step 7
To Do
Simulate the time-domain performance

Ziegler-Nichols Method

The Ziegler-Nichols rule is a heuristic PID tuning rule that attempts to produce good values for the three PID gain parameters.

Step 1

Terms	Procedure(s)	Notes
• K_p: the controller path gain. • T_i: the controller's integrator time constant. • T_d: the controller's derivative time constant.	1. Set $T_i=\infty$. 2. Set $T_d=0$. 3. $K_p=0$.	The PID controller is a P controller now.

Step 2

Terms	Procedure(s)	Notes
• K_u: ultimate (or critical) gain	1. Increase K_p until there are sustained oscillations in the signals in the control system. 2. This K_p value is denoted as K_u.	As shown in the figure, when constant amplitude oscillation obtained, K_u was selected.

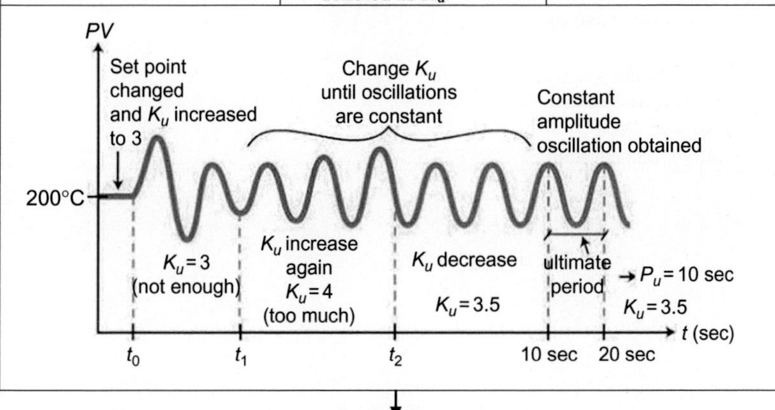

Step 3

Terms	Procedure(s)	Notes
• P_u: ultimate (or critical) period (s).	Measure P_u of the sustained oscillations.	P_u is the time to the signal to start having constant oscillation after reaching K_u.

A

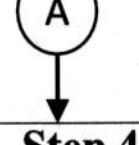

Step 4			
Procedure(s)			
Based on the controller type, the controller parameter values can be calculated according to the following table.			
Control Type	**K_p**	**T_i**	**T_d**
P	0.5 K_u	-	-
PI	0.45 K_u	P_u/ 1.2	-
PD	0.8 K_u	-	P_u/8
Classic PID	0.6 K_u	P_u/2	P_u/8
Pessen integral rule	0.7 K_u	P_u/2.5	P_u3/20
Some overshoot	0.33 K_u	P_u/2	P_u/3
No overshoot	0.2 K_u	P_u/2	P_u/3

Step 5	
Terms	**Procedure(s)**
• u(t): The control signal. • K_p: The controller path gain. • e(t): Difference between the current value and the set point. • T_i: the controller's integrator time constant. • T_d: the controller's derivative time constant.	Use the values of the parameter which was calculated in the previous step and put in the following equation. $u(t) = K_p\left(e(t) + \frac{1}{T_i}\int_0^t e(t')dt' + T_d\frac{de(t)}{dt}\right)$

Chapter 15

Design and Simulation Softwares

ABSTRACT

The computer programs can help a mechatronic system designer in many different forms, from designing to simulating the system. Mechanical and circuit drawing programs are added here again to provide complete knowledge. Finite element analysis programs help the designer to analyze the system before physically building it, while CAM (Computer-Aided Manufacturing) programs make it possible to generate g-codes to be used by CNC machines.

When the mechatronics system software is being designed, the designer should take into account both the hardware and the software for better performance. The basic parts of the software, such as interrupts, should be known in detail for better control. Meanwhile the hardware, the microcontroller properties, should support the software. Standard communication protocol should also be used for the standardization of the software, or else no other programmer will be able to understand the project and the system cannot be integrated to other systems.

Mechatronic Components. https://doi.org/10.1016/B978-0-12-814126-7.00015-3

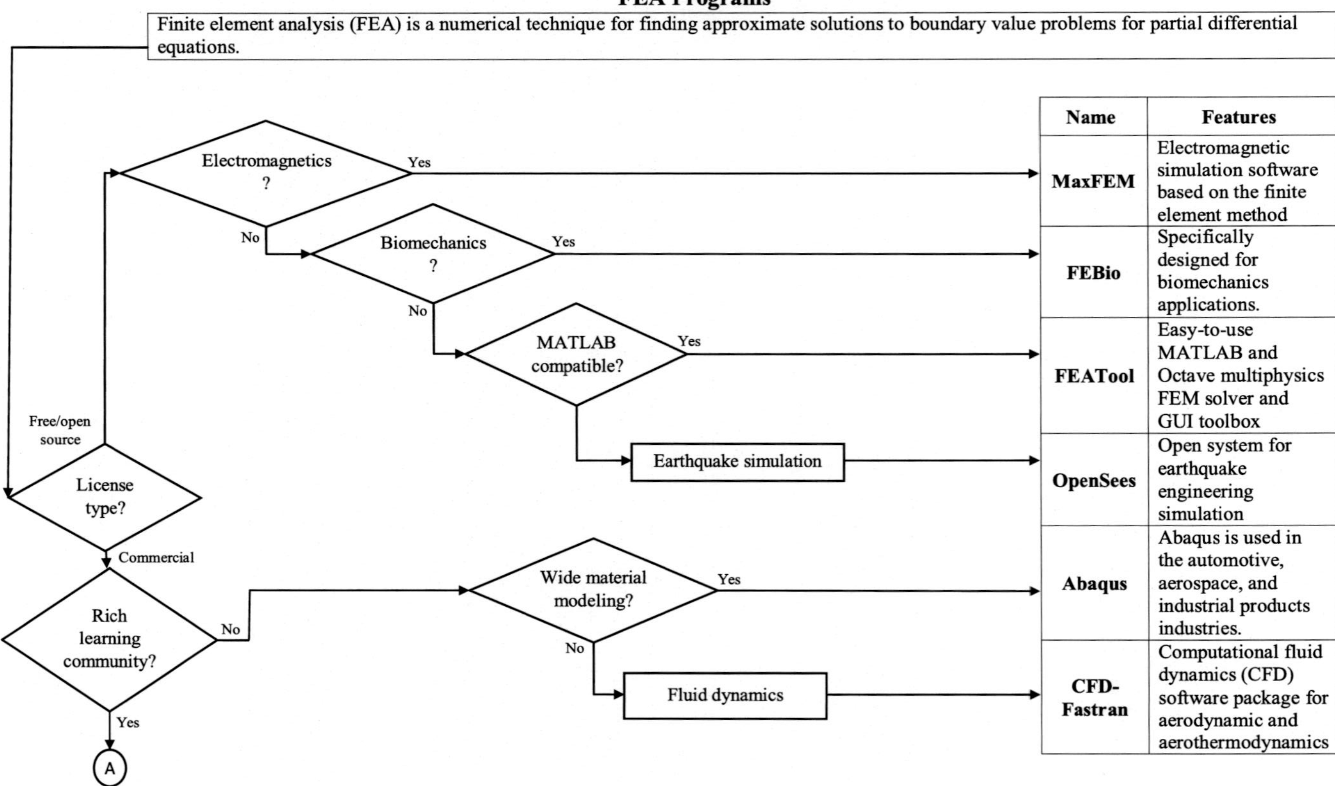

Name	Features
MaxFEM	Electromagnetic simulation software based on the finite element method
FEBio	Specifically designed for biomechanics applications.
FEATool	Easy-to-use MATLAB and Octave multiphysics FEM solver and GUI toolbox
OpenSees	Open system for earthquake engineering simulation
Abaqus	Abaqus is used in the automotive, aerospace, and industrial products industries.
CFD-Fastran	Computational fluid dynamics (CFD) software package for aerodynamic and aerothermodynamics

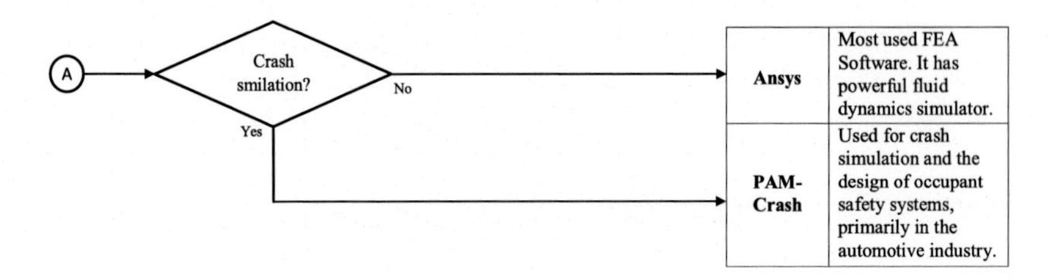

CAM Programs

Computer-aided manufacturing (CAM) is the use of computer software to control machine tools and related machinery in the manufacturing of workpieces.

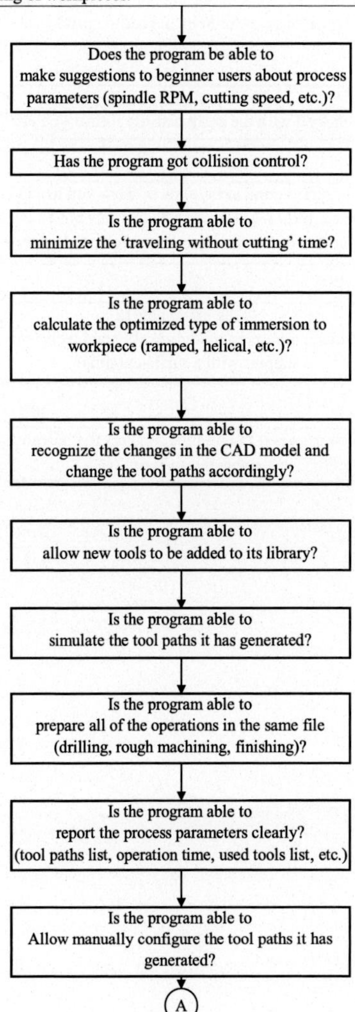

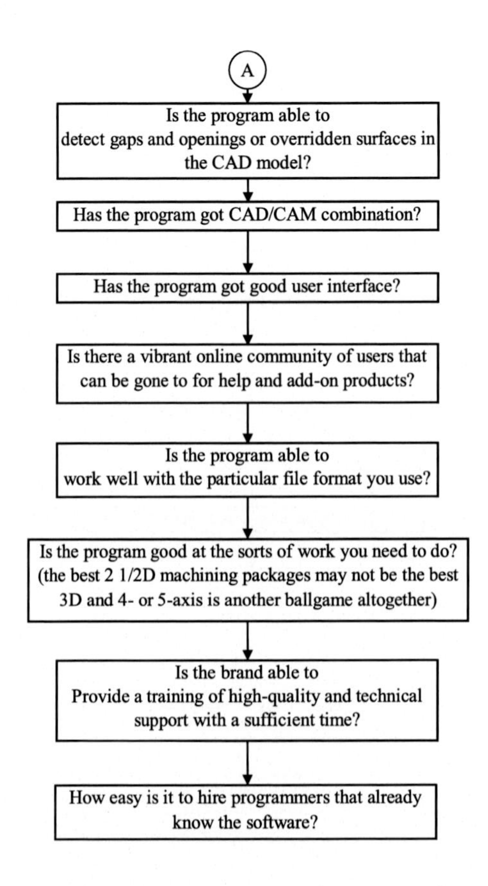
A
Is the program able to
detect gaps and openings or overridden surfaces in
the CAD model?
Has the program got CAD/CAM combination?
Has the program got good user interface?
Is there a vibrant online community of users that
can be gone to for help and add-on products?
Is the program able to
work well with the particular file format you use?
Is the program good at the sorts of work you need to do?
(the best 2 1/2D machining packages may not be the best
3D and 4- or 5-axis is another ballgame altogether)
Is the brand able to
Provide a training of high-quality and technical
support with a sufficient time?
How easy is it to hire programmers that already
know the software?

Drawing Programs

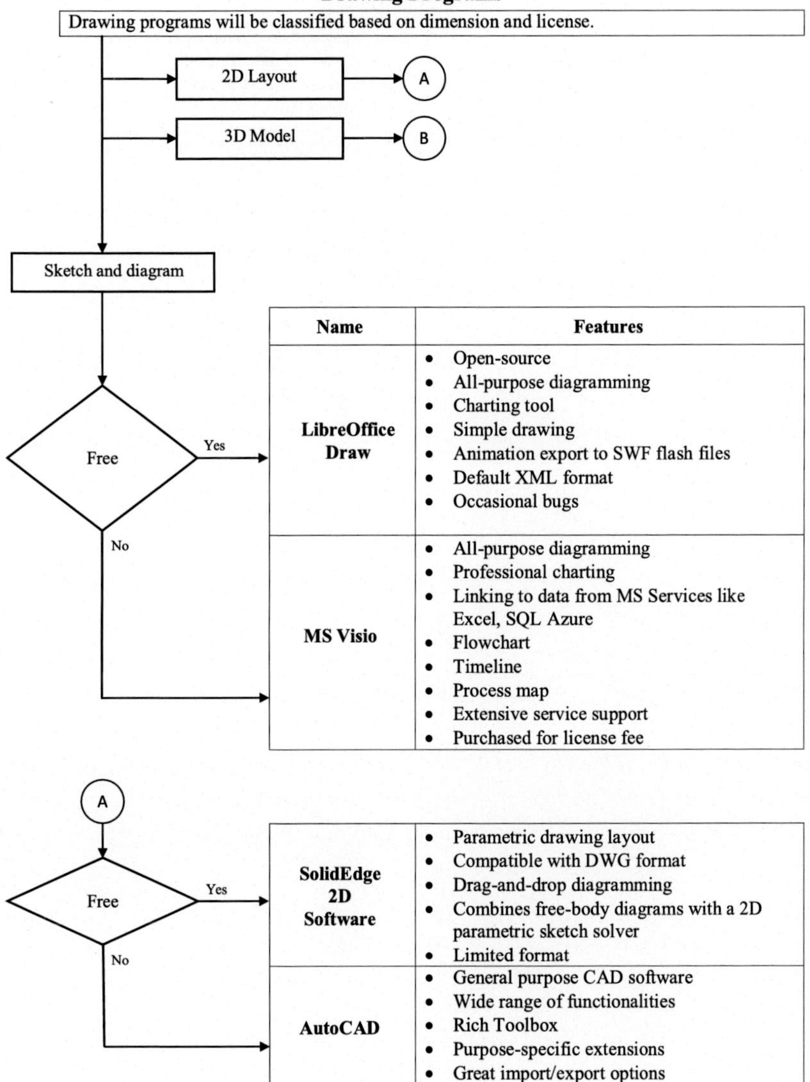

Name	Features
LibreOffice Draw	• Open-source • All-purpose diagramming • Charting tool • Simple drawing • Animation export to SWF flash files • Default XML format • Occasional bugs
MS Visio	• All-purpose diagramming • Professional charting • Linking to data from MS Services like Excel, SQL Azure • Flowchart • Timeline • Process map • Extensive service support • Purchased for license fee

SolidEdge 2D Software	• Parametric drawing layout • Compatible with DWG format • Drag-and-drop diagramming • Combines free-body diagrams with a 2D parametric sketch solver • Limited format
AutoCAD	• General purpose CAD software • Wide range of functionalities • Rich Toolbox • Purpose-specific extensions • Great import/export options

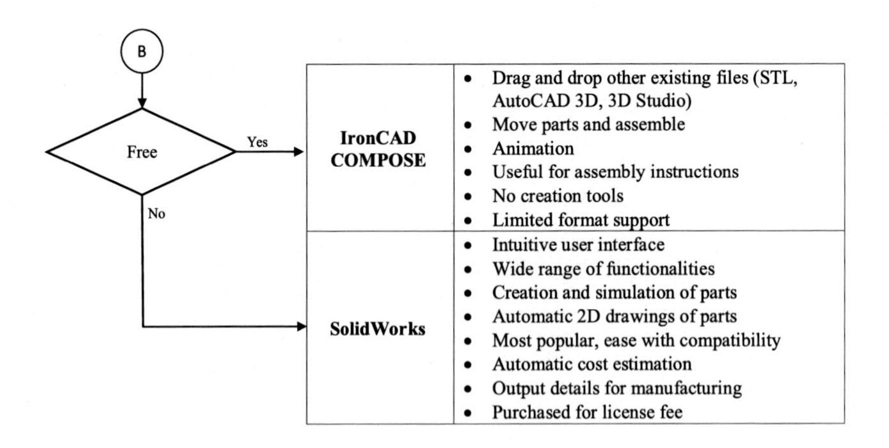
B
Free
Yes
No
IronCAD COMPOSE
Drag and drop other existing files (STL, AutoCAD 3D, 3D Studio)
Move parts and assemble
Animation
Useful for assembly instructions
No creation tools
Limited format support
SolidWorks
Intuitive user interface
Wide range of functionalities
Creation and simulation of parts
Automatic 2D drawings of parts
Most popular, ease with compatibility
Automatic cost estimation
Output details for manufacturing
Purchased for license fee

Meshing

The partial differential equations that govern a physical phenomenon (fluid flow, heat transfer, stress distribution, etc.) are not usually amenable to analytical solutions, except for very simple cases. Therefore, in order to analyze a physical phenomenon, corresponding physical domains are split into smaller subdomains (cells or elements). The governing equations are then discretized and solved inside each of these subdomains and then put together to get a complete solution of the phenomenon inside the domain.

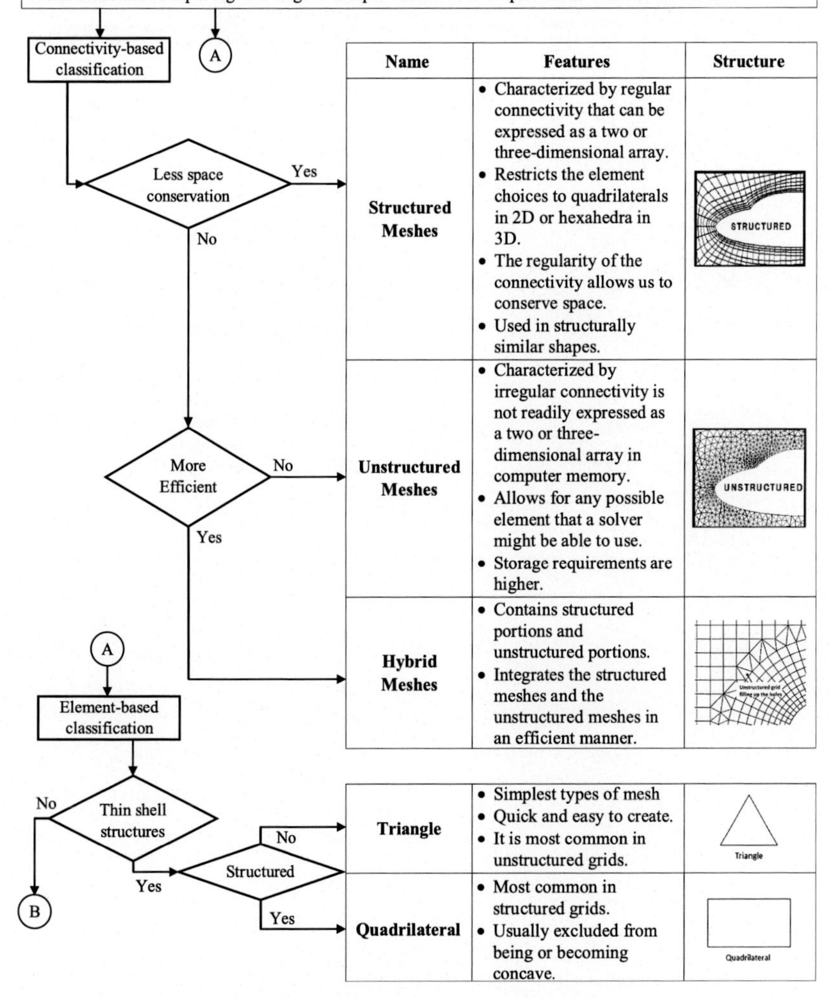

Name	Features	Structure
Structured Meshes	• Characterized by regular connectivity that can be expressed as a two or three-dimensional array. • Restricts the element choices to quadrilaterals in 2D or hexahedra in 3D. • The regularity of the connectivity allows us to conserve space. • Used in structurally similar shapes.	
Unstructured Meshes	• Characterized by irregular connectivity is not readily expressed as a two or three-dimensional array in computer memory. • Allows for any possible element that a solver might be able to use. • Storage requirements are higher.	
Hybrid Meshes	• Contains structured portions and unstructured portions. • Integrates the structured meshes and the unstructured meshes in an efficient manner.	

Name	Features	Structure
Triangle	• Simplest types of mesh • Quick and easy to create. • It is most common in unstructured grids.	
Quadrilateral	• Most common in structured grids. • Usually excluded from being or becoming concave.	

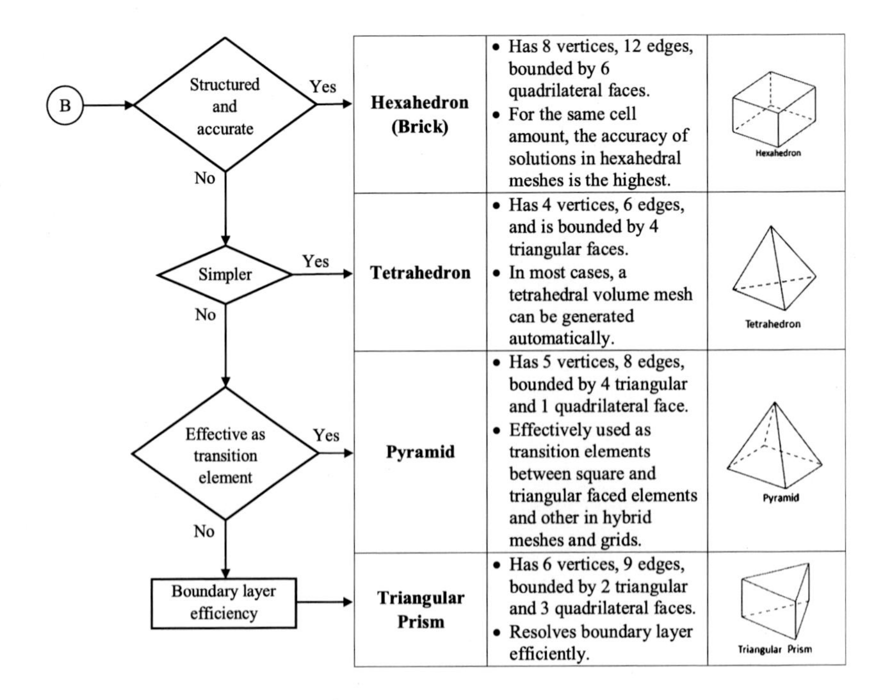
B
Structured and accurate
Yes
No
Hexahedron (Brick)
• Has 8 vertices, 12 edges, bounded by 6 quadrilateral faces.
• For the same cell amount, the accuracy of solutions in hexahedral meshes is the highest.
Hexahedron
Simpler
Yes
No
Tetrahedron
• Has 4 vertices, 6 edges, and is bounded by 4 triangular faces.
• In most cases, a tetrahedral volume mesh can be generated automatically.
Tetrahedron
Effective as transition element
Yes
No
Pyramid
• Has 5 vertices, 8 edges, bounded by 4 triangular and 1 quadrilateral face.
• Effectively used as transition elements between square and triangular faced elements and other in hybrid meshes and grids.
Pyramid
Boundary layer efficiency
Triangular Prism
• Has 6 vertices, 9 edges, bounded by 2 triangular and 3 quadrilateral faces.
• Resolves boundary layer efficiently.
Triangular Prism

Circuit Drawing Programs

Circuit drawing programs will be classified based on the ease of use and professionalism.

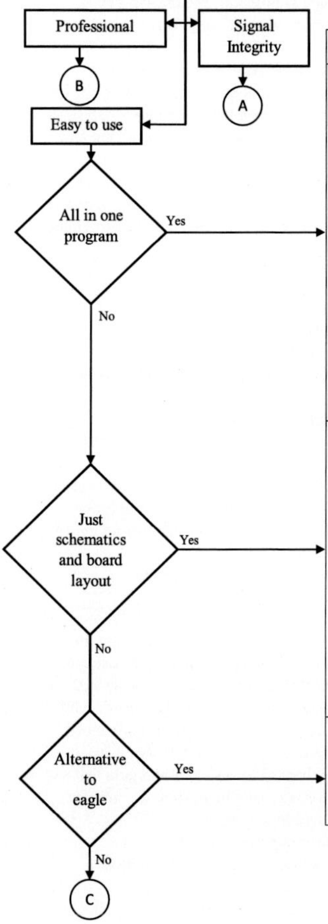

Professional → B

Signal Integrity → A

Easy to use

All in one program — Yes → Proteus; No ↓

Just schematics and board layout — Yes → Eagle; No ↓

Alternative to eagle — Yes → KiCAD; No → C

Name	Features
Proteus	• Incorporates many functions derived from several other languages: C, BASIC, Assembly, Clipper/dBase. • Student usually uses this program. • Have extensive library of components; you can try several different components that you do not have on hand. • The integrated package with common user interface and the context-sensible help can make the learning process more quick and easy. • Virtual prototype with Proteus VSM reduce the time and cost of software and hardware development. • Faster and easier to connect the components; not losing in tangles of wires. • Does not prefer in real-time applications.
Eagle	• Provides a fast learning curve, even for those new to PCB design. • Its extensive and fully open component libraries ease the design process for all. • Growing in its capabilities and workflow compatibility, demonstrated by the hundreds of extensions (ULPs) openly available to all users and its structured XML file format. • Schematics and board layout design software only, no practical simulation. • No maintenance fees and the licensing is really flexible. • Easy to use, easy to learn. •
KiCAD	• Also free and open source. • Just schematic, no simulation tool as eagle. • People use as alternative to eagle. • Has good interface, PCB view is also good, everything well labeled.

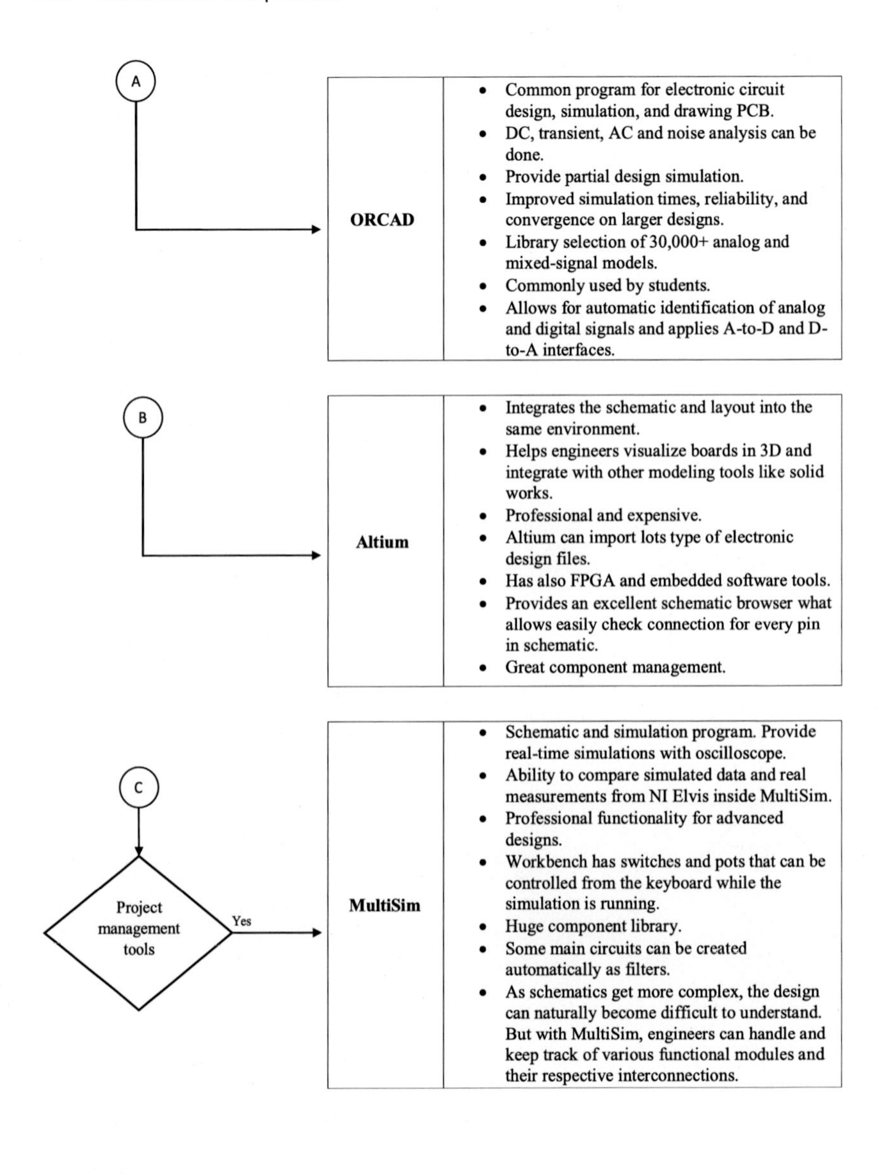
A
ORCAD
Common program for electronic circuit design, simulation, and drawing PCB.
DC, transient, AC and noise analysis can be done.
Provide partial design simulation.
Improved simulation times, reliability, and convergence on larger designs.
Library selection of 30,000+ analog and mixed-signal models.
Commonly used by students.
Allows for automatic identification of analog and digital signals and applies A-to-D and D-to-A interfaces.
B
Altium
Integrates the schematic and layout into the same environment.
Helps engineers visualize boards in 3D and integrate with other modeling tools like solid works.
Professional and expensive.
Altium can import lots type of electronic design files.
Has also FPGA and embedded software tools.
Provides an excellent schematic browser what allows easily check connection for every pin in schematic.
Great component management.
C
Project management tools
Yes
MultiSim
Schematic and simulation program. Provide real-time simulations with oscilloscope.
Ability to compare simulated data and real measurements from NI Elvis inside MultiSim.
Professional functionality for advanced designs.
Workbench has switches and pots that can be controlled from the keyboard while the simulation is running.
Huge component library.
Some main circuits can be created automatically as filters.
As schematics get more complex, the design can naturally become difficult to understand. But with MultiSim, engineers can handle and keep track of various functional modules and their respective interconnections.

Software Parts

Coding activity starts when the design of the software has been done and the specifications of the module to be developed are available

↓

Step 1		
To do	**Language**	**Features**
Writing the code to implement functionalities in the software	HTML/CSS	Essential for creating static Web pages
	JavaScript	Popular, multipurpose, provides behavior of Web, i.e., a field in the browser turns red
	Java	Widely used for anything from Web applications to desktop and mobile apps
	Objective-C	Can be used to write desktop applications
	Swift	Can be used to compute Apple iOS Applications
	PHP	Used by heavy hitters, i.e., Facebook, for coding for the Web
	Python	Used for everything from server automation to data science
	Ruby	Was used for Twitter. Used widely among Web startups

↓

Step 2 ← B		
To do	**Software Part**	**Features**
Implementing the software parts according to the project needs	Functions (group of statements)	Functions consist of statements that tell the computer, what to do (i.e., get user input, display output, set values, do arithmetic, etc.)
	Libraries	• Program Libraries: Collection of (usually) precompiled, reusable programming routines, that programmers can "call" while programming =) no need for writing the code again • Storage Library: Collection of physical storage media, such as tapes, discs and a way to access them • Data Library: Area of data center, where storage media are archived • Virtual Library: Online version of a traditional library
	Variable	Value that can change, depending on conditions or on information passed through the program
	Interrupt	Signal from a program within the system that causes operating system to stop and then continue to run the current program or start another one
	Syntax	Specific grammar in the computer language
	Semantics	Tells what the things mean
	Loop	Sequence of instructions, which is continually repeated until certain condition is reached

↓

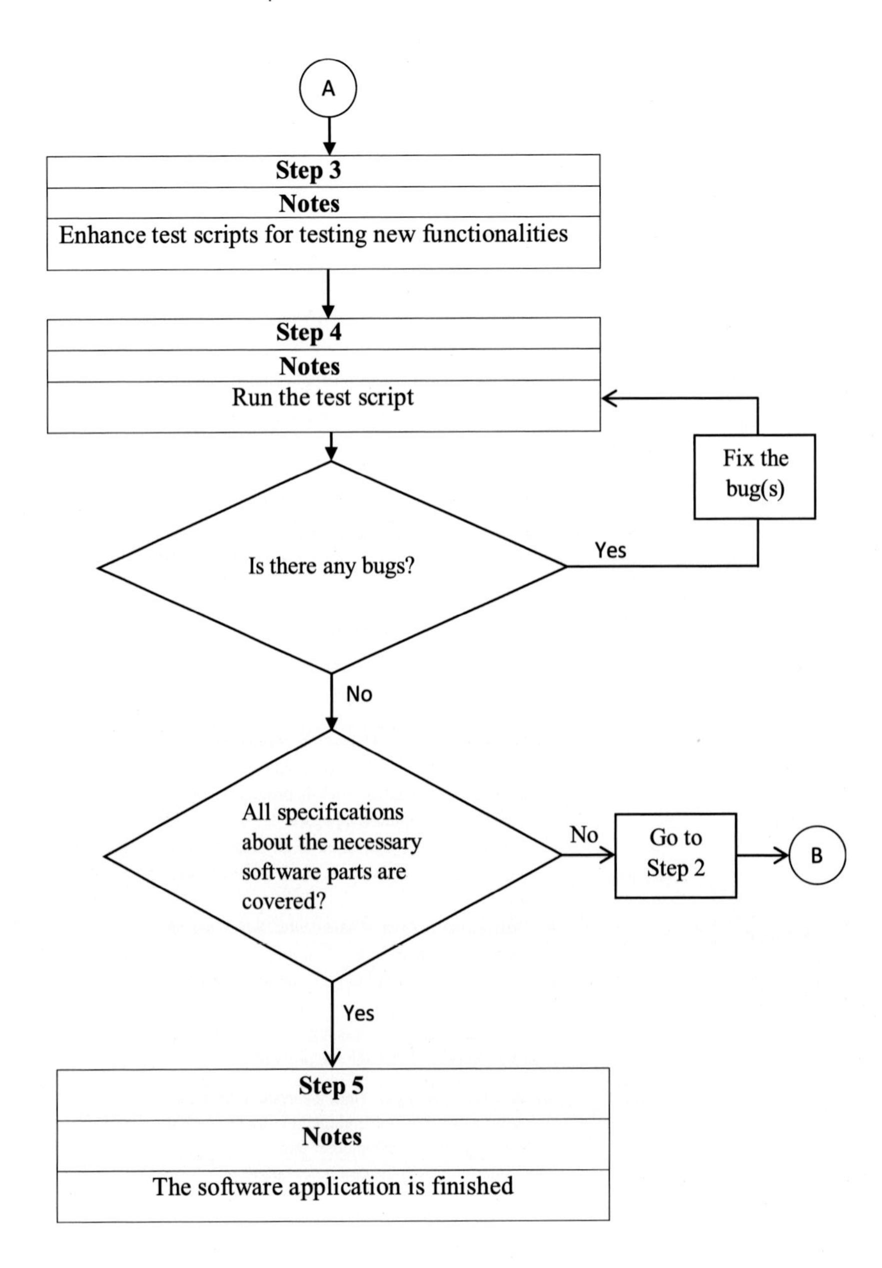
A
Step 3
Notes
Enhance test scripts for testing new functionalities
Step 4
Notes
Run the test script
Fix the bug(s)
Yes
Is there any bugs?
No
All specifications about the necessary software parts are covered?
No
Go to Step 2
B
Yes
Step 5
Notes
The software application is finished

Interrupts

Interrupt is an event external to the currently executing process that causes a change in the normal flow of instruction execution. (Classification has done in general case however the inner properties from the 8086 processor)

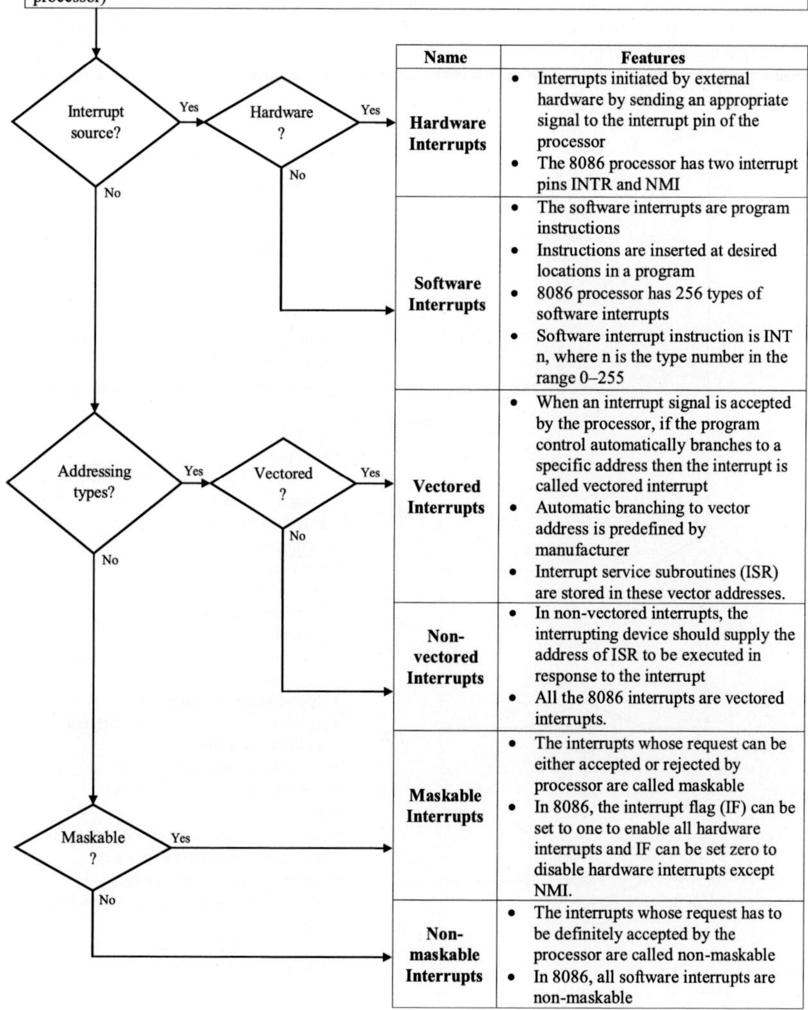

Name	Features
Hardware Interrupts	• Interrupts initiated by external hardware by sending an appropriate signal to the interrupt pin of the processor • The 8086 processor has two interrupt pins INTR and NMI
Software Interrupts	• The software interrupts are program instructions • Instructions are inserted at desired locations in a program • 8086 processor has 256 types of software interrupts • Software interrupt instruction is INT n, where n is the type number in the range 0–255
Vectored Interrupts	• When an interrupt signal is accepted by the processor, if the program control automatically branches to a specific address then the interrupt is called vectored interrupt • Automatic branching to vector address is predefined by manufacturer • Interrupt service subroutines (ISR) are stored in these vector addresses.
Non-vectored Interrupts	• In non-vectored interrupts, the interrupting device should supply the address of ISR to be executed in response to the interrupt • All the 8086 interrupts are vectored interrupts.
Maskable Interrupts	• The interrupts whose request can be either accepted or rejected by processor are called maskable • In 8086, the interrupt flag (IF) can be set to one to enable all hardware interrupts and IF can be set zero to disable hardware interrupts except NMI.
Non-maskable Interrupts	• The interrupts whose request has to be definitely accepted by the processor are called non-maskable • In 8086, all software interrupts are non-maskable

Communication Protocols

A communication protocol is a system of rules that allow two or more entities of a communications system to transmit information via any kind of variation of a physical quantity.
Communication protocols consist of several layers. All of these layers have certain rules in order to work properly. Lower layers define how a signal is transmitted through a transport medium while higher layers define how the devices will behave during communication and how to interpret incoming signal packets. Some protocols cover all of these layers while others may cover certain layers which allow them to be used with other different protocols that covers those layers.

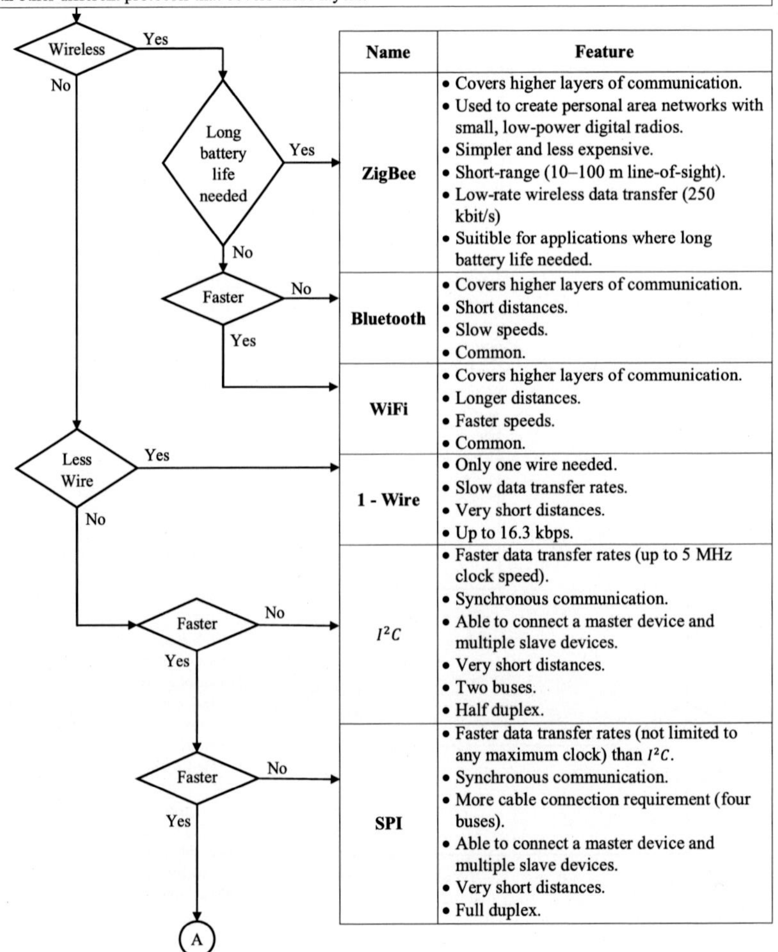

Name	Feature
ZigBee	• Covers higher layers of communication. • Used to create personal area networks with small, low-power digital radios. • Simpler and less expensive. • Short-range (10–100 m line-of-sight). • Low-rate wireless data transfer (250 kbit/s) • Suitible for applications where long battery life needed.
Bluetooth	• Covers higher layers of communication. • Short distances. • Slow speeds. • Common.
WiFi	• Covers higher layers of communication. • Longer distances. • Faster speeds. • Common.
1 - Wire	• Only one wire needed. • Slow data transfer rates. • Very short distances. • Up to 16.3 kbps.
I^2C	• Faster data transfer rates (up to 5 MHz clock speed). • Synchronous communication. • Able to connect a master device and multiple slave devices. • Very short distances. • Two buses. • Half duplex.
SPI	• Faster data transfer rates (not limited to any maximum clock) than I^2C. • Synchronous communication. • More cable connection requirement (four buses). • Able to connect a master device and multiple slave devices. • Very short distances. • Full duplex.

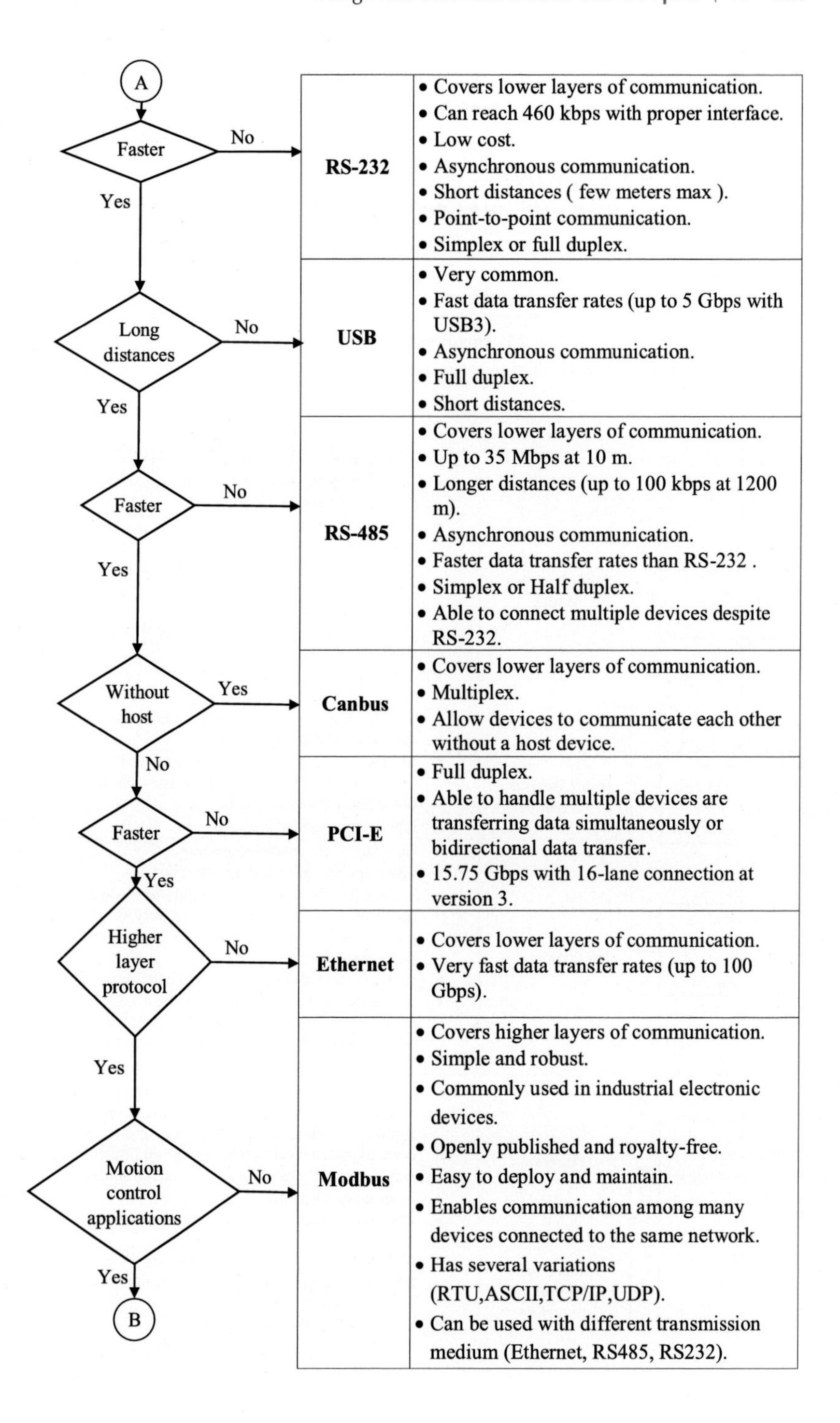
A
Faster
No
Yes
RS-232
• Covers lower layers of communication.
• Can reach 460 kbps with proper interface.
• Low cost.
• Asynchronous communication.
• Short distances (few meters max).
• Point-to-point communication.
• Simplex or full duplex.
Long distances
No
Yes
USB
• Very common.
• Fast data transfer rates (up to 5 Gbps with USB3).
• Asynchronous communication.
• Full duplex.
• Short distances.
Faster
No
Yes
RS-485
• Covers lower layers of communication.
• Up to 35 Mbps at 10 m.
• Longer distances (up to 100 kbps at 1200 m).
• Asynchronous communication.
• Faster data transfer rates than RS-232 .
• Simplex or Half duplex.
• Able to connect multiple devices despite RS-232.
Without host
Yes
No
Canbus
• Covers lower layers of communication.
• Multiplex.
• Allow devices to communicate each other without a host device.
Faster
No
Yes
PCI-E
• Full duplex.
• Able to handle multiple devices are transferring data simultaneously or bidirectional data transfer.
• 15.75 Gbps with 16-lane connection at version 3.
Higher layer protocol
No
Yes
Ethernet
• Covers lower layers of communication.
• Very fast data transfer rates (up to 100 Gbps).
Motion control applications
No
Yes
B
Modbus
• Covers higher layers of communication.
• Simple and robust.
• Commonly used in industrial electronic devices.
• Openly published and royalty-free.
• Easy to deploy and maintain.
• Enables communication among many devices connected to the same network.
• Has several variations (RTU,ASCII,TCP/IP,UDP).
• Can be used with different transmission medium (Ethernet, RS485, RS232).

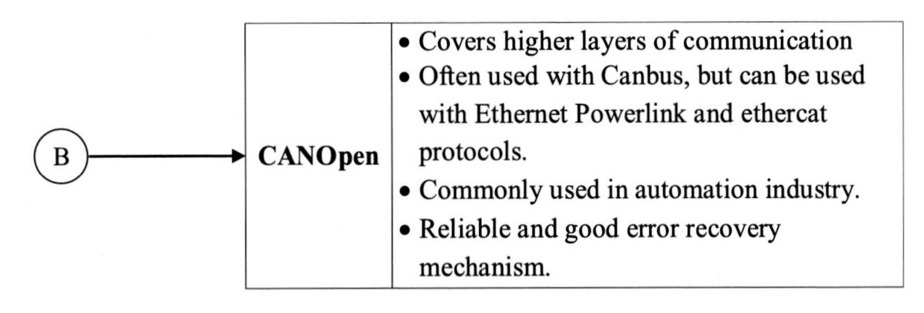

Microcontroller Properties

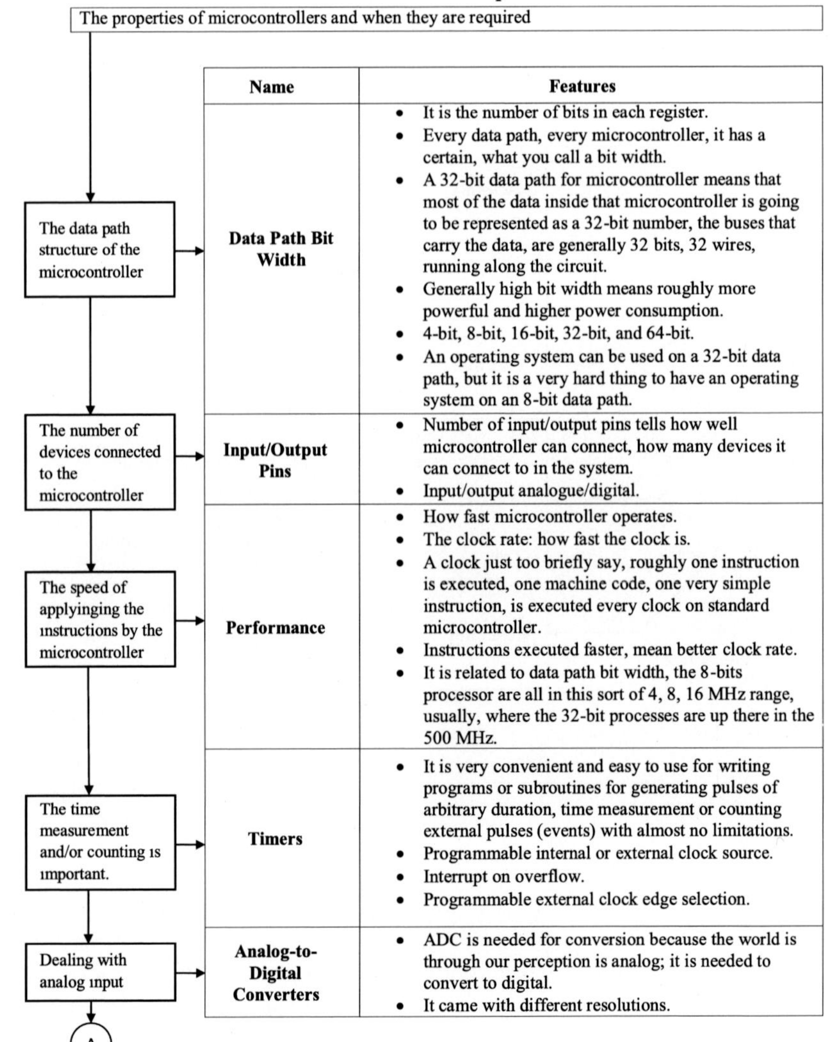

Name	Features
Data Path Bit Width	• It is the number of bits in each register. • Every data path, every microcontroller, it has a certain, what you call a bit width. • A 32-bit data path for microcontroller means that most of the data inside that microcontroller is going to be represented as a 32-bit number, the buses that carry the data, are generally 32 bits, 32 wires, running along the circuit. • Generally high bit width means roughly more powerful and higher power consumption. • 4-bit, 8-bit, 16-bit, 32-bit, and 64-bit. • An operating system can be used on a 32-bit data path, but it is a very hard thing to have an operating system on an 8-bit data path.
Input/Output Pins	• Number of input/output pins tells how well microcontroller can connect, how many devices it can connect to in the system. • Input/output analogue/digital.
Performance	• How fast microcontroller operates. • The clock rate: how fast the clock is. • A clock just too briefly say, roughly one instruction is executed, one machine code, one very simple instruction, is executed every clock on standard microcontroller. • Instructions executed faster, mean better clock rate. • It is related to data path bit width, the 8-bits processor are all in this sort of 4, 8, 16 MHz range, usually, where the 32-bit processes are up there in the 500 MHz.
Timers	• It is very convenient and easy to use for writing programs or subroutines for generating pulses of arbitrary duration, time measurement or counting external pulses (events) with almost no limitations. • Programmable internal or external clock source. • Interrupt on overflow. • Programmable external clock edge selection.
Analog-to-Digital Converters	• ADC is needed for conversion because the world is through our perception is analog; it is needed to convert to digital. • It came with different resolutions.

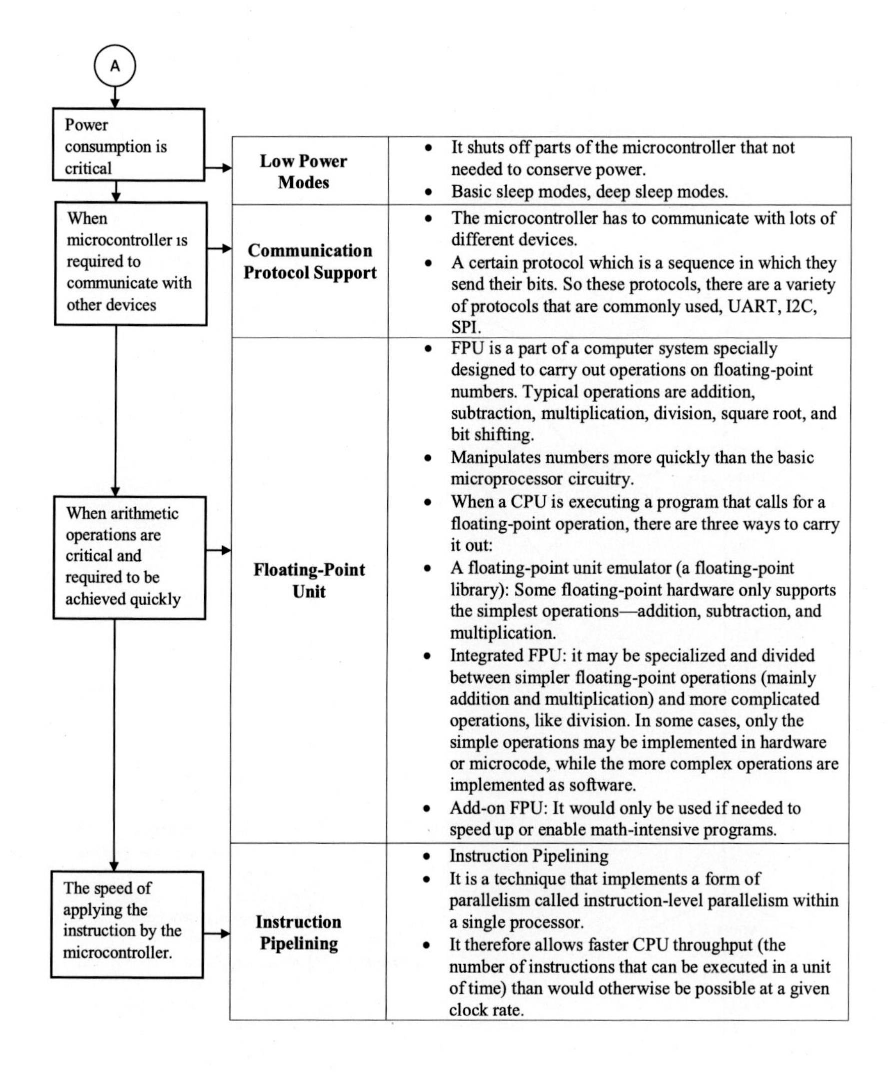
A
Power consumption is critical
Low Power Modes
It shuts off parts of the microcontroller that not needed to conserve power.
Basic sleep modes, deep sleep modes.
When microcontroller is required to communicate with other devices
Communication Protocol Support
The microcontroller has to communicate with lots of different devices.
A certain protocol which is a sequence in which they send their bits. So these protocols, there are a variety of protocols that are commonly used, UART, I2C, SPI.
When arithmetic operations are critical and required to be achieved quickly
Floating-Point Unit
FPU is a part of a computer system specially designed to carry out operations on floating-point numbers. Typical operations are addition, subtraction, multiplication, division, square root, and bit shifting.
Manipulates numbers more quickly than the basic microprocessor circuitry.
When a CPU is executing a program that calls for a floating-point operation, there are three ways to carry it out:
A floating-point unit emulator (a floating-point library): Some floating-point hardware only supports the simplest operations—addition, subtraction, and multiplication.
Integrated FPU: it may be specialized and divided between simpler floating-point operations (mainly addition and multiplication) and more complicated operations, like division. In some cases, only the simple operations may be implemented in hardware or microcode, while the more complex operations are implemented as software.
Add-on FPU: It would only be used if needed to speed up or enable math-intensive programs.
The speed of applying the instruction by the microcontroller.
Instruction Pipelining
Instruction Pipelining
It is a technique that implements a form of parallelism called instruction-level parallelism within a single processor.
It therefore allows faster CPU throughput (the number of instructions that can be executed in a unit of time) than would otherwise be possible at a given clock rate.

Controllers

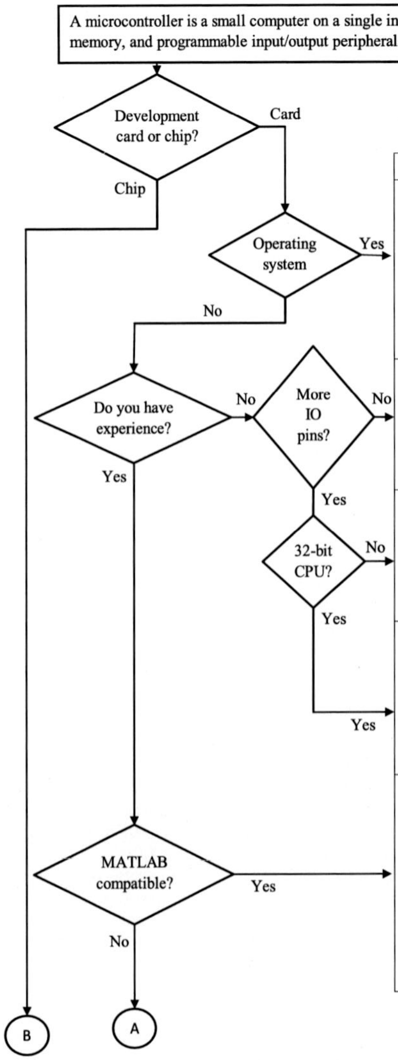

A microcontroller is a small computer on a single integrated circuit containing a processor core, memory, and programmable input/output peripherals.

Development card or chip?
Card
Chip
Operating system
Yes
No
Do you have experience?
No
More IO pins?
No
Yes
Yes
32-bit CPU?
No
Yes
Yes
MATLAB compatible?
Yes
No
B
A

Name	Features
Raspberry Pi	• Linux operating system • GPIO • HDMI • SD Card • USB • UART • Ethernet • Good for software engineers
Arduino Uno	• 14 digital IO pins • 6 analog pins • USB • UART • Good for beginners • Wide community
Arduino Mega	• 54 digital IO pins • 16 analog pins • USB • UART • Wide community • Good for beginners
Arduino Due	• 54 digital IO pins • 16 analog pins • USB • UART • Wide community • 32-bit ARM Core • Good for beginners
STM32F4 Discovery	• 32-bit ARM Cortex M4 core • USB • UART • On-board debugger • Accelerometer • Microphone • Speaker driver • Floating-point accelerator. • Good for control and DSP applications

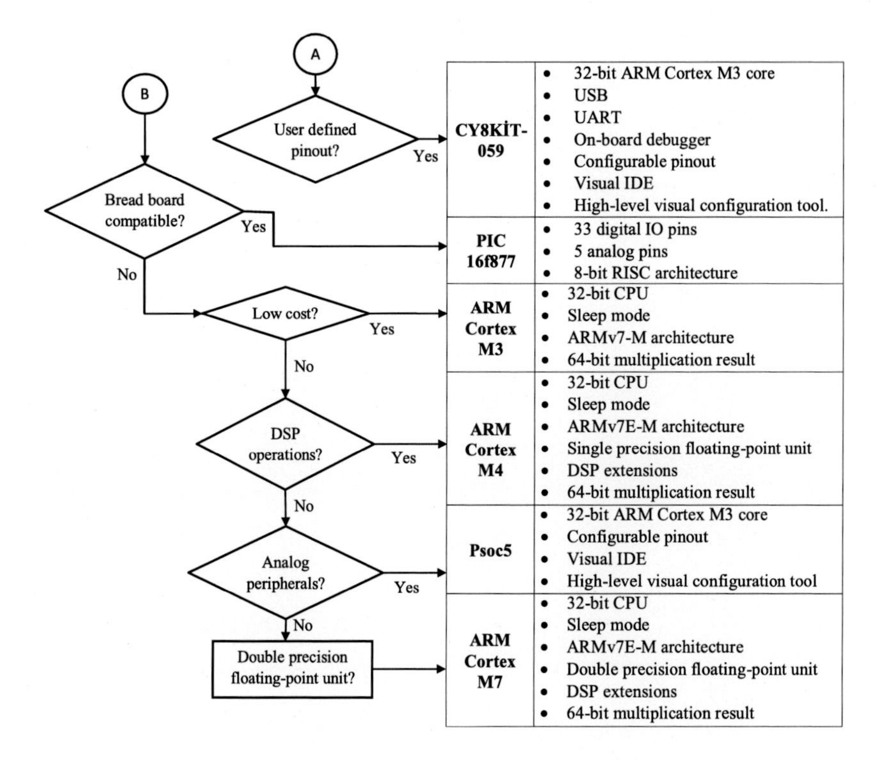
A
B
User defined pinout?
Yes
Bread board compatible?
Yes
No
Low cost?
Yes
No
DSP operations?
Yes
No
Analog peripherals?
Yes
No
Double precision floating-point unit?
CY8KİT-059
32-bit ARM Cortex M3 core
USB
UART
On-board debugger
Configurable pinout
Visual IDE
High-level visual configuration tool.
PIC 16f877
33 digital IO pins
5 analog pins
8-bit RISC architecture
ARM Cortex M3
32-bit CPU
Sleep mode
ARMv7-M architecture
64-bit multiplication result
ARM Cortex M4
32-bit CPU
Sleep mode
ARMv7E-M architecture
Single precision floating-point unit
DSP extensions
64-bit multiplication result
Psoc5
32-bit ARM Cortex M3 core
Configurable pinout
Visual IDE
High-level visual configuration tool
ARM Cortex M7
32-bit CPU
Sleep mode
ARMv7E-M architecture
Double precision floating-point unit
DSP extensions
64-bit multiplication result

Chapter 16

Case Studies

When classic engineering education is considered, education is given by teaching the theory of the subjects followed by projects and hands on experience. The fundamental problem in mechatronics education is that the field covers mechanical, electrical, and computer engineering fields and it is too broad. Moreover, the real problem lies in the fact that the systems to be designed are complex and a decision made in one part of the system effects the other parts. In order to overcome this difficulty, this book can be used to understand and compare the advantages and disadvantages of the subsystems. The following examples show how this book can be used for mechatronic system design.

Case 1: Flow of material inside 3D printer nozzle

The importance of 3D printers in engineering design is increasing remarkably. Some of the methods a decade ago were impossible to do because of the cost but now are both affordable and relatively easy.

During a nose surgery the doctors use a suction system to drain the bleeding. In this project, the team was asked to design a plastic pipe to be put inside the nasal cavity to drain the blood. First of all a mold is made and by using silicon a pipe is manufactured. It would be very difficult to make a mold without a three-dimensional (3D) printer.

In the second stage of the project the team realized that it would be very advantageous to make a 3D printer, which can use a biomaterial. Even though the earlier pipe could be used in the surgery, some patients might have an allergic reaction to silicon.

When the 3D bioprinter is being designed, the team faced the problem of the material flowing inside the extruder. In Chapter 3, Reynolds number is explained, which defines if the flow is laminar or turbulent. Since the material is heated from outside of the nozzle, mixing of the material is desired for better heat distribution. In order to increase the turbulence, the Reynolds number should be bigger than 4000. There are different ways to increase the Reynolds number, based on head drop, pressure drop, and flow coefficient. By using these different formulas the designer can change the extruder design to ensure turbulent flow.

Mechatronic Components. https://doi.org/10.1016/B978-0-12-814126-7.00016-5

Case 2: How to choose the right material for a lightweight throwable robot

Military applications of mobile robotics are gaining more popularity and the design of robots with more payload capacity is very important in the battlefield. The weight of the robot can be divided into two parts: the weight of the machine and the payload capacity. When a mobile robot is being designed if the weight of the machine can be kept lower, the payload capacity can be higher.

In order to be able to make the robot lightweight, different types of materials should be studied (Chapter 5). Considering that the robot will be working in the field, the material should be corrosion resistant. Considering the density and strength of the material, aluminum and titanium are good choices. When the 7000 series Aluminum is examined, the designer can see that strength to weight ratio is very high and this material is very good for lightweight design.

Case 3: Manufacturing process selection

Unfortunately, so many engineers think that designing a machine is drawing. In reality, the manufacturing of the part is also very important and it is easy to design a part that is difficult and expensive to manufacture. The technical drawing programs are very advanced. In Chapter 8, the drawing programs are summarized. The two-dimensional (2D) softwares have been replaced by 3D ones over the years but the 2D softwares are still very good for layout designs and are used by civil and industrial engineers. For manufacturing purposes, 3D programs are more useful in that these designs can be used by CNC machines. The designer can generate the necessary g-codes to manufacture the part; however, during this process manufacturing methods, cutting tools, and the material properties should be considered.

For example, if the part is cubical rather than cylindrical, it is mostly manufactured by a milling operation. If the number of parts is high enough, a cost analysis should be done, and the parts can be manufactured by molding. If the strength and fatigue resistance of a part are important, forging will be a better choice than casting, but it will be a more expensive procedure. The manufacturing process depends on the part shape, material of the part, and cost analysis.

The assembly of the device is the next step after manufacturing and the parts should be designed for ease of assembly as well.

In order to design a device for ease of assembly, the number of parts should be minimized, parts should be self-locating and fastening, and top-down assembly process should be used. Using standard parts with modular and symmetrical design also helps.

Case 4: Power transmission

Almost all the mechatronic systems have some kind of motion inside them. There are different ways to generate motion with especially electrically powered motors (explained in the following case studies). After the motion is generated, most of the time the speed and torque of the motor should be transferred to another point inside the machine. There are so many choices available for power transmission (Chapter 7).

During the design of a snow-blowing robot, a gasoline engine is used for high-energy output. The energy generated by the gasoline engine is used in two points: at the blades of the snow blower and at the alternator to generate electricity.

The snow blower robot has mechanical blades and a pump to take the snow from the front and to throw it out to the side. The transmitting of power from the engine to the blades and pump was achieved by a chain because of the high torques. On the other hand, the alternator was powered via a belt system since it is cheaper than a chain system.

The hybrid system was built by using a gasoline engine and an alternator. The robot was powered via 4 direct current (DC) motors and the energy to these motors was supplied by the alternator. By using DC motors, there was no need for a differential at the back wheels of the robot and a steering mechanism in the front wheels.

Case 5: Which cables to use?

Any mechatronic application will require cabling. In practice, a cable can be used for two different purposes, to carry power or signal. At first, the designer should know the operating voltage and current as well as maximum allowed voltage and current for the surge duration. Installation conditions also should be known prior to cable selection. By using the necessary formulas (Chapter 10), the load on the cable can be calculated. If the calculation is not done, the design can be either overrated, causing a ticker cable to be used, or because of the use of too thin cables, the wire will heat up and the insulation over the cable will get damaged and will cause electrical failure in the system. Circuit breakers and fuses can be used to decrease the damage of any electrical failure.

Case 6: Actuator selection and motor drivers

Electrical motors are preferred in mechatronics applications because of their ease of control. A gasoline powered motor, a hydraulic, and a pneumatic actuator have higher inertia compared to electrical motors and nonlinearity characteristic in input to output.

During the selection of the motor (electrical actuator) the task should be specified. In some applications, like an industrial robotic arm, the position and the speed of the motion is important. In some applications, like running a conveyer belt, only motion generation is necessary. Since the alternating current (AC) voltage is very easy to have inside a building and relatively cheaper, an AC motor is the perfect choice for inside application where only motion is necessary.

Motor drivers are necessary because the control electronics run with low power and low current (Chapter 10). For high-voltage and high-current motors, a motor driver is necessary to control the electrical energy flowing from the energy source to the motor. Motor drivers are selected depending on the motor type. It is not possible to drive a stepper motor with a DC motor driver. Voltage and current level of the driver is selected depending on the motor input power. Lower current values will cause the driver to burn. When the system is not working, the driver still consumes some energy. This can become very important if the machine is working in the field and has a lot of waiting time because of the nature of the task; for example, a border patrolling robot. The motor drivers can support different types of control inputs, such as pulse width modulation (PWM), which allows the motor to be rotated at a desired speed; however, the designer should remember the torque drop in PWM drives. Finally, the number of channels of the motor driver becomes important allowing multiple number of motors to be driven.

Case 7: Microcontroller selection

The development boards come in different sizes (Chapter 15) with varying computing capacity and input/output ratio. The application dictates which board should be used. For a robotic application, when a board is chosen, the following parameters are taken into consideration: processor speed, number of pins, memory size, programming language, and price. If the application requires heavy computing processes to be done in a short period of time, such as using a Kalman filter on the sensor data, the clock speed of the processor should have a high priority. If the code sketch is significantly long and has a relatively larger size in bytes, memory size should be considered. Code sketch size becomes relatively big, if there are "if" and "for" commands in the program. The programming language can differ with the preferences of the coder, if the coder is more familiar with Arduino than C++ or Python; the language becomes a deterministic factor. The number of I/O pins required on the board can change with the application, and more intricate designs may need a higher number of pins on the board. If high computing power is required, a board using an ARM process can be chosen, since it has enough computing power to achieve even image processing on board. In simple applications, an Arduino board is selected for its small size, sufficient capacity, low cost, and market availability.

Case 8: Encoder selection

If the mechatronic application requires the system to be positioned and/or speed controlled, the direction, the amount, and the speed of the rotation should be measured. An encoder is a sensor that measures direction and amount of rotation and with calculation the speed of rotation can be found. From the structural perspective there are two types of encoders: absolute and incremental. An absolute encoder can provide output for the specific position and if the power is cut, the system does not forget where it was when the power comes back. An Incremental encoder has a disc with multiple cuts and a light source and light detector can count the number of cuts to find the amount of rotation. In order to calculate the speed of the rotation the amount of rotation is divided into time intervals. If the time is too long (depends on the application), this division will not be able to catch any change in the speed and will just average it. If the time interval taken is too short, it causes computational load.

Index

Note: Page numbers followed by *f* indicate figures.

Printed in the United States
By Bookmasters